Sustainability of Green and Eco-friendly Composites

The book explores the pertinent aspects of sustainability of green and eco-friendly composites including their development methods and processing, characterization, properties, and applications. Significance for the design and engineering of high-performance green and eco-friendly composites is discussed in the present book. Insights on a wide spectrum of potential advanced applications ranging from automotive and aerospace to biomedical and packaging, etc. using these are highlighted. Further, it discusses life cycle and carbon footprint assessment of sustainable materials.

Features:

- Explores different processing methods of green and eco-friendly composites.
- Discusses development and optimization of green nanocomposites for sustainable manufacturing.
- Collates modern green and eco-friendly composites research from theory to application.
- Covers hybridization of reinforced fibers on the performance of green and eco-friendly composites.
- Analyzes and discusses calculation of carbon footprint and Life Cycle Assessment of composites.

This book is aimed at graduate students and researchers in materials science and engineering, sustainable materials, composites, and nanomaterials.

Emerging Materials and Technologies

Series Editor: Boris I. Kharissov

The Emerging Materials and Technologies series is devoted to highlighting publications centered on emerging advanced materials and novel technologies. Attention is paid to those newly discovered or applied materials with potential to solve pressing societal problems and improve quality of life, corresponding to environmental protection, medicine, communications, energy, transportation, advanced manufacturing, and related areas.

The series takes into account that, under present strong demands for energy, material, and cost savings, as well as heavy contamination problems and worldwide pandemic conditions, the area of emerging materials and related scalable technologies is a highly interdisciplinary field, with the need for researchers, professionals, and academics across the spectrum of engineering and technological disciplines. The main objective of this book series is to attract more attention to these materials and technologies and invite conversation among the international R&D community.

Advanced Functional Metal-Organic Frameworks
Fundamentals and Applications
Edited by Jay Singh, Nidhi Goel, Ranjana Verma and Ravindra Pratap Singh

Nanoparticles in Diagnosis, Drug Delivery and Nanotherapeutics
Edited by Divya Bajpai Tripathy, Anjali Gupta, Arvind Kumar Jain, Anuradha Mishra and Kuldeep Singh

Functional Nanomaterials for Sensors
Suresh Sagadevan and Won-Chun Oh

Functional Biomaterials
Advances in Design and Biomedical Applications
Anuj Kumar, Durgalakshmi Dhinasekaran, Irina Savina, and Sung Soo Han

Smart Nanomaterials
Imalka Munaweera and M. L. Chamalki Madhusha

Nanocosmetics
Drug Delivery Approaches, Applications and Regulatory Aspects
Edited by Prashant Kesharwani and Sunil Kumar Dubey

Sustainability of Green and Eco-friendly Composites
Edited by Sumit Gupta, Vijay Chaudhary, and Pallav Gupta

For more information about this series, please visit: www.routledge.com/Emerging-Materials-and-Technologies/book-series/CRCEMT

Sustainability of Green and Eco-friendly Composites

Edited by Sumit Gupta, Vijay Chaudhary, and Pallav Gupta

CRC Press
Taylor & Francis Group
Boca Raton London New York

CRC Press is an imprint of the
Taylor & Francis Group, an **informa** business

First edition published 2024
by CRC Press
6000 Broken Sound Parkway NW, Suite 300, Boca Raton, FL 33487–2742

and by CRC Press
4 Park Square, Milton Park, Abingdon, Oxon, OX14 4RN

CRC Press is an imprint of Taylor & Francis Group, LLC

© 2024 selection and editorial matter, Sumit Gupta, Vijay Chaudhary, and Pallav Gupta; individual chapters, the contributors

ISBN: 9781032224527 (hbk)
ISBN: 9781032224534 (pbk)
ISBN: 9781003272625 (ebk)
DOI: 10.1201/9781003272625

Typeset in Times
by Apex CoVantage, LLC

Contents

Preface

The present edited book volume named *Sustainability of Green and Eco-friendly Composites* will provide a systematic knowledge of various aspects of green and eco-friendly composites with special emphasis on sustainability. This edited book will cover various methods to develop green and eco-friendly composites along with their characterization techniques to study the various properties like mechanical, tribological, thermal, electrical etc. It also provides insight on a wide spectrum of potential advanced applications ranging from automotive and aerospace, to biomedical and packaging, etc. using green and eco-friendly composites which will also be highlighted in the present volume.

The book will also address the advantages of natural fibers to promote sustainability and the eco-friendly nature of developed composites. Life cycle assessment is a prominent tool to analyze the carbon footprints of composites, which leads to the development of high-quality eco-friendly composites. In short, this book will focus on the synthesis, characterization and application of green and eco-friendly composites with its sustainability.

The contents of this book will be beneficial for students who are attending various programs in mechanical engineering, civil engineering, material science, chemistry, physics as well as researchers working both in industry and academia.

The present book consists of 14 chapters whose details are as appended next:

The first chapter discusses in detail the development and mechanical characterization of light weight green composites for sustainable development. In this chapter the different properties of natural fiber composites are also illustrated.

The second chapter is dedicated to the fabrication and characterization of green composites. Apart from that, the potential applications of green composites in petroleum products are also discussed.

The third chapter discusses the effect of processing on natural fibers for composite manufacturing. Apart from that, the various physical and chemical properties are also discussed in this chapter.

The fourth chapter covers the mechanical characterization of Labeo Catla and Laevistrombus Canarium Derived Hydroxyapatite-High Density Polyethylene Composite. For the prepared composites the various mechanical properties are reported in this chapter.

The fifth chapter describes the fabrication and characterization of lead-free magnetic-ferroelectric green composites for spintronic applications. Various structural and multiferroic properties are discussed in this chapter for the fabricated composites.

The sixth chapter is focused on the hybridization of reinforced fibers on the performance of green and eco-friendly composites. The different types of fibers like synthetic fibers, kevlar fibers, glass fibers and carbon fibers are discussed in detail in this chapter.

The seventh chapter explains about the fatigue phenomenon in natural fiber composites. The different phenomenon of fatigue analysis and design has been discussed in this chapter. Apart from that, theory of fatigue damage evolution is also discussed in detail.

The eighth chapter discusses the finite element method in damage modelling of green composites. Representative volume method (RVE) and homogenization which is emerged out to be the most popular technique for evaluating the various properties of natural fiber composites is also discussed in this chapter.

The ninth chapter discusses the lead-free multiferroic $BiFeO_3$ based sustainable green composites. The various topics covered in this chapter are applications, opportunities and future challenges.

The tenth chapter focuses on the sustainable green composites for packaging applications. In this chapter the different criterions are also discussed for the packaging applications.

The eleventh chapter discusses the sustainable green composites for structural applications along with its characteristics. The various structural and mechanical properties are also discussed in this chapter.

The twelfth chapter illustrates the synthesis of ionic polymer metal composites for robotic application. The different techniques for the preparation of ionic polymer metal composites are also discussed in this chapter.

The thirteenth chapter discusses the carbon footprint analysis of green composites. Carbon footprint and its effect are specifically discussed in this chapter.

The fourteenth chapter illustrates the life cycle assessment of eco-friendly composites. Different aspects of life cycle assessment for green and eco-friendly composites are discussed in detail in this chapter.

All fourteen chapters given by different authors are written in a systematic manner thereby covering all the major topics of green and eco-friendly composites.

All three editors are thankful to almighty God. Apart from this, Dr. Sumit Gupta and Dr. Vijay Chaudhary are also thankful to their family members for the support extended during the editing of this book. Dr. Pallav Gupta is also thankful to his mother (Beena Gupta), wife (Dr. Ritu Agrahari) and son (Saahas Gupta) for the encouragement as well as support extended during the entire duration of editing this book.

We editors are also thankful to all our contributors who submitted their chapters in the present book.

We are also thankful to Dr. Gagandeep Singh and Ms. Aditi Mittal along with their entire team of CRC Press (Taylor & Francis Group) for publishing this book in the fastest possible time and in the most efficient manner.

Dr. Sumit Gupta
Dr. Vijay Chaudhary
Dr. Pallav Gupta

Editors Biographies

Dr. Sumit Gupta is presently working as an Assistant Professor (Grade-III) in the Department of Mechanical Engineering, Amity School of Engineering and Technology, Amity University Uttar Pradesh, Noida, India. Prior to this he has also served as an Assistant Professor, School of Engineering, G. D. Goenka University, Gurugram, India. He graduated in mechanical engineering from University of Rajasthan in 2008 and earned master's as well as a doctorate degree from Malaviya National Institute of Technology-Jaipur, India in 2010 and 2016 respectively. His areas of research are sustainable manufacturing, lean manufacturing, reverse logistics, sustainable product design and sustainable supply chain management. Dr. Gupta has over 10 years of teaching and research experience. He has published over 70 research papers in peer reviewed international journals as well as in reputed international and national conferences. A large number of students have completed their summer internships and B.Tech. Projects under his guidance. He has guided 10 M.Tech. Dissertations and is presently guiding 2 Ph.D. scholars as well. He is Reviewer of various national and international journals. He is a member of various international and national professional societies.

Dr. Vijay Chaudhary is currently working as an Assistant Professor (Grade-II) in the Department of Mechanical Engineering, Amity School of Engineering and Technology (A.S.E.T.), Amity University Uttar Pradesh, Noida (INDIA). He has completed his B.Tech. in 2011 from the Department of Mechanical Engineering, Uttar Pradesh Technical University, Lucknow, India and then completed M.Tech. (Hons) in 2013 from the Department of Mechanical Engineering, Madan Mohan Malviya Engineering College, Gorakhpur, India. He has completed his Ph.D. in 2019 from Department of Mechanical Engineering, Netaji Subhas University of Technology, University of Delhi, India. His research area of interest lies in the processing and characterization of polymer composites, tribological analysis of bio-fiber based polymer composites, water absorption of bio-fiber based polymer composites, and surface modification techniques related to polymer composite materials. Dr. Chaudhary has over 8 years of teaching and research experience. He has published more than 60 research papers in peer reviewed international journals as well as in reputed international and national conferences. He has published 16 book chapters with reputed publishers. More than 25 students have completed their summer internships, B.Tech. Projects and M.Tech. Dissertations under his guidance. At present he is also guiding several Ph.D. scholars. Currently, he is working in the field of bio-composites, nano-composites and smart materials.

Dr. Pallav Gupta is presently working as an Assistant Professor (Grade-III) in the Department of Mechanical Engineering, Amity School of Engineering and Technology, Amity University Uttar Pradesh, Noida (INDIA). He completed his B.Tech. (Honors) from the Department of Mechanical Engineering, Integral University, Lucknow, INDIA in the year 2009, Qualified GATE in 2009 with AIR-3291 and then completed his M.Tech. (Honors) from I.I.T. (B.H.U.), INDIA in the year 2011 followed by a Ph.D. in the year 2015 from I.I.T. (B.H.U.), INDIA. His area of research includes material processing, composite materials, mechanical behaviour and corrosion. Dr. Gupta has over 8 years of teaching and research experience. He has published over 100 research papers in peer reviewed international journals as well as in reputed international and national conferences in India as well as in abroad. Apart from this he has also published 12 chapters in books published by Springer, Elsevier and Taylor & Francis. Dr. Gupta has edited 06 books and has also authored 2 textbooks namely on (a) *Manufacturing Processes* and (b) *Principles of Management*. Many students have completed their Summer Internships, B.Tech. Projects and M.Tech. Dissertations under his guidance. 03 scholars have completed and 05 are presently registered for their Ph.D. research work under his supervision/co-supervision in the area of "Coatings, Metal Matrix Composites and Polymer Matrix Composites".

Contributors

S. Amsamani
Department of Mechanical Engineering, Coimbatore Institute of Technology, Coimbatore

Ankit
Department of Mechanical Engineering, Government Engineering College Jhalawar, Rajasthan-326023, India

Vishal Arya
Department of Mechanical Engineering, I.E.T. Bundelkhand University, Jhansi, India

C. Balaji Ayyanar
Department of Mechanical Engineering, Coimbatore Institute of Technology, Coimbatore—641 014, Tamil Nādu, India

C. Bharathira
HAL, Bangalore

Anurag Dixit
Department of Mechanical and Automation Engineering, G.B Pant Government Engineering College, New Delhi, India 1100202

Brijesh Gangil
Department of Mechanical Engineering, H.N.B. Garhwal University, Srinagar Garhwal, India

Tannu Garg
Department of Applied Physics, AIAS, Amity University, Noida (U.P.) 201303, India

S. Gaurav
Department of Applied Physics, AIAS, Amity University, Noida (U.P.) 201303, India

B. Gayathri
Department of Chemistry, Coimbatore Institute of Technology

Ravi Kumar Goyal
Mechanical Engineering Department, Nirwan University, Jaipur, India

Ashutosh Gupta
Department of Mechanical Engineering, G. B. Pant Institute of Engineering and Technology, Pauri, Garhwal, Uttarakhand, India

Tanika Gupta
Department of Applied Physics, AIAS, Amity University, Noida (U.P.) 201303, India

Yusuf Jameel
Advanced Engineering Materials and Composites Research Centre, Department of Mechanical and Manufacturing Engineering, Faculty of Engineering, University Putra Malaysia, 43400 UPM Serdang, Selangor, Malaysia.

Mohammad Zain Khan
Department of Chemistry, Aligarh Muslim University

Z. R. Khan
Department of Physics, College of Science, University of Hai'l, P. O. Box-2440, Hai'l, Saudi Arabia

Shara Khursheed
Department of Mechanical Engineering, Integral University Lucknow

Arvind Kumar
Experimental Research Laboratory, Department of Physics, ARSD College, University of Delhi, New Delhi-110021, India

Manish Kumar
Experimental Research Laboratory,
Department of Physics, ARSD
College, University of Delhi,
New Delhi-110021, India

Mukesh Kumar
Department of Mechanical and
Automation Engineering,
G.B Pant Government Engineering
College, New Delhi, India 1100202

Sandeep Kumar
Department of Mechanical Engineering,
H.N.B. Garhwal University,
Srinagar Garhwal, India

Satyam Kumar
Department of Physics and Electronics,
Hansh Raj College, University of
Delhi, New Delhi-110007, India

Shivani A. Kumar
Amity Institute of Applied Sciences,
Amity University Uttar Pradesh,
Noida, 201313

K. Marimuthu
Department of Mechanical
Engineering, Coimbatore Institute of
Technology, Coimbatore—641 014,
Tamilnadu, India

K. M. Moeed
Department of Mechanical Engineering,
Integral University Lucknow

S. K. Pradeep Mohan
Department of Mechanical
Engineering, Coimbatore Institute of
Technology, Coimbatore—641 014,
Tamil Nādu, India

Himanshu Prajapati
Department of Mechanical Engineering,
Netaji Subhas University of
Technology, New Delhi-10078 India

K. Pratibha
Department of Applied Physics, AIAS,
Amity University, Noida (U.P.)
201303, India

Lalit Ranakoti
Mechanical Engineering Department,
Graphic Era Deemed to be
University, Dehradun, India

Moti Lal Rinawa
Department of Mechanical Engineering,
Government Engineering College
Jhalawar, Rajasthan-326023, India

S. Shankar
Experimental Research Laboratory,
Department of Physics, ARSD
College, University of Delhi, Dhaula
Kuan, New Delhi-110021, India

Gaurav Sharma
Amity Institute of Applied Sciences,
Amity University Uttar Pradesh,
Noida-201313, India

Subhash Sharma
CONACyT- Centro de Nanociencias
y Nanotecnología, Universidad
NacionalAutónoma de, México, Km
107 Carretera Tijuana-Ensenada s/n,
Ensenada, B.C., C.P. 22800, México

Sushil Kumar Singh
Mechanical Engineering Department,
Nirwan University, Jaipur, India

Abhay Nanda Srivastava
Department of Chemistry, Nitishwar
Mahavidyalaya, A constituent
Unit of B.R.A. Bihar University,
Muzaffarpur-842002, India

Anshuman Srivastava
Department of Mechanical Engineering,
Shambhunath Institute of Engineering
and Technology, Prayagraj, India

R. Sunitha
Department of Textiles and Clothing,
Avinashilingam Institute for Home
Science and Higher Education for
Women, Coimbatore

Abhishek Tevatia
Department of Mechanical Engineering,
Netaji Subhas University of
Technology, New Delhi, India 110078

N. Vasugi
Department of Textiles and Clothing,
Avinashilingam Institute for Home
Science and Higher Education for
Women, Coimbatore

Jitendra Verma
Department of Mechanical
Engineering, I.E.T. Bundelkhand
University, Jhansi, India

Rohit Verma
Department of Applied Physics,
AIAS, Amity University,
Noida (U.P.) 201303,
India

Shashikant Verma
Department of Mechanical
Engineering,
I.E.T. Bundelkhand University,
Jhansi, India

1 Development and Mechanical Characterization of Light Weight Green Composites for Sustainable Development

Sushil Kumar Singh, Yusuf Jameel,
Ravi Kumar Goyal, and Anshuman Srivastava

CONTENTS

DOI: 10.1201/9781003272625-1

1.1   INTRODUCTION

Composites are made up of two materials, one of which is known as the reinforcing phase and comes in the form of fibers, sheets, or particles, and the other is known as the matrix phase and is implanted in the reinforcing phase (Charlet, 2007). Plastic is a synthetic-based polymer composite that is almost universally used. Natural fibers are increasingly being used as a result of growing environmental awareness and concern for environmental sustainability. Lingo cellulose is found in natural fibers, which are derived from plants. A natural fiber filament is about 10µm long, with three secondary cell walls and a primary cell wall. Jute, kenaf, flax, sisal, hemp, and oil palm fruit are just a few examples of natural fibers. Natural fibers are durable, environmentally friendly, light weight, renewable, inexpensive, and biodegradable. The micro fibrils are made up of 30–100 cellulose molecules and have a diameter of about 10 nm. Natural fiber is used to reinforce thermosetting and thermoplastic matrices. The most commonly used thermosetting resins are epoxy, polyester, polyurethane, and phenolic. They have high mechanical properties like strength and stiffness, which make them ideal for high-performance applications. Because of their low cost, natural fibers are used in construction, packaging, railway coach interiors, and the automobile industry. Natural fiber's main disadvantage is its hydrophilic nature, which has an impact on mechanical properties such as flexural strength and fracture toughness (Raja et al., 2017).

1.2   GREENER ALTERNATIVE MATERIALS

Seeing as natural fibers have numerous advantages, including low cost, light weight, and environmental friendliness, researchers have begun to focus more attention on this area in order to maximize the benefits of using natural fibers as reinforcement in polymers (Al-oqla et al., 2014; Bledzki et al., 2004). Natural fibers are used to make a variety of composites for a range of applications. Fibers outperform glass in many ways, including environmental impacts during manufacturing operations and environmental pollution after use of the product, and the light weight of natural fiber composites makes them suitable for automobile applications, increasing fuel efficiency and lowering emissions to the environment (Al-oqla & Sapuan, 2014). In comparison to metals, fiber reinforced polymeric composites have a high specific strength and modulus. Most composites are currently made with non-biodegradable polymeric resins like epoxies and polyurethane, or high-strength fibers like graphite, aramids, and glass. Petroleum is used to make many of these polymers and fibers, which is a non-renewable resource. As a result, using composites instead of common plastics has improved product performance while also reducing weight and cost. Composites are difficult to dispose of or recycle because they are made up of two or more materials (Al-oqla et al., n.d.).

1.3   PROPERTIES OF NATURAL FIBER COMPOSITES

1.3.1   Moisture Absorption

Hydrophilic in nature, all cellulosic fibers absorb moisture from the environment until equilibrium is reached. Natural fibers regain moisture at a rate ranging from 5% to 12%, as shown in Table 1.1. Moisture absorption affects the interface and mechanical

TABLE 1.1
Comparison between the Properties of Different Reinforcements

Plant fibers	Tensile strength (MPa)	Young modulus (GPa)	Specific modulus (GPa)	Tenacity (MN/m²)	Density (g/cm³)	Moisture regain(%)
Cotton	400–700	6–10	4–6.5	–	1.55	8.51
Kapok	93.21	41	12.9	–	0.45	10.9
Bamboo	571	27	18	–	1.52	–
Flax	510–910	50–70	34–48	–	1.45	12
Hemp	300–760	30–60	20–41	–	1.43	12
Jute	200–460	20–55	14–39	440–553	1.34	12
Kenaf	300–1200	22–60	–	–	1.301	17
Ramie	9151	231	15	–	1.551	8.51
Abaca	141	411	–	–	1.52	141
Banana	530	27–32	20–24	529–754	1.35	–
Pine apple	414	60–82	42–57	413–162	1.44	–
Sisal	100–800	9–22	6–15	568–640	1.45	11
Coir	100–200	6	5.2	131–175	1.15	13

Source: Dubrovski and Golob (2009)

properties of composite materials by causing dimensional variations in the fiber and composite material. Moisture can cause a poor fiber–matrix interface during composite fabrication. Because of the poor interface, load transfer will be ineffective, and the composite material will deteriorate (Taj & Munawar, 2014). Natural fibers may be treated with various chemicals or grafted with vinyl monomers to help reduce moisture regain (Panigrahi et al., 2007).

1.4 FACTORS AFFECTING MECHANICAL PERFORMANCE OF NATURAL FIBER COMPOSITE

1.4.1 FIBER SELECTION

Plant, animal, and mineral origins are the most common classifications for fiber types. Plant fiber is mostly made up of cellulose, whereas animal fiber is mostly made up of protein (Shah et al., 2014). In most cases, geographic factors relating to fiber availability play a significant role in fiber selection (Fadzli et al., 2015). Higher cellulose content and more aligned cellulose microfibrils in the fiber direction results in higher structural requirements, which improves performance. Natural fiber properties vary significantly depending on chemical composition and structure, which are related to fiber type, extraction method, harvesting time, treatment, and storage procedures. The strength of flax fibers extracted manually has been shown to be 20% greater than those extracted manually after five days, and the strength of mechanically extracted flax fibers has been found to be 15% lower (Pickering et al., 2007). The young's modulus significantly reduces with moisture content and increases with temperature; fiber strength increases with moisture content and decreases with temperature (Charlet, 2007). Fibers are normally stronger and more stable than the matrix, and the composite's strength and stiffness increase as the fiber content increases. This is ultimately dependent on the fiber/matrix interfacial strength, which can be reduced with high hydrophobic matrices such as polypropylene with increasing fiber content until additives or other methods are used, regardless of young's modulus, which increases with fiber content but more gradually than when the interface is optimized (Beg, 2007). When a rational interfacial strength is established, the strength of injection molded thermoplastic matrix composites peaks at 40–55%, with lower contents due to poor wetting, which results in reduced stress transfer across the fiber-matrix interface and increased porosity. Due to less reliance on interfacial strength than composite strength, stiffness can be increased up to higher fiber contents of 55–65 m% with similar materials (Le & Pickering, 2015). Increased fiber length also increases fiber load bearing efficiency; however, if the fiber length is too long, the fiber may become tangled, resulting in poor fiber dispersion, lowering overall reinforcement efficiency (Devi et al., 1996).

1.4.2 MATRIX SELECTION

On a fiber-reinforced composite, the matrix is crucial. It protects the fibers' surfaces from mechanical damage, serves as a barrier against the elements, and transfers loads to the fibers. Polymeric matrices are the most commonly used because they are light weight and can be processed at low temperatures. With natural fiber, both thermoplastic and thermoset polymers are used as matrices (Holbery & Houston, 2006). The temperature at which the fibers degrade limits matrix selection. Natural fibers used as reinforcement are thermally unstable above 200°C, though they can be processed at higher temperatures for a short period of time in some circumstances (Summerscales, 2021). Only thermoplastics and thermosets, such as polyethylene, polyolefin, PP, polyvinyl chloride, and polystyrene, are used as a matrix due to this limitation, as they soften below this temperature (Aparecido & Giriolli, n.d.). Epoxy resin, polyester, phenol formaldehyde, and VE resins are among the thermosets used. Thermoplastics are softened repeatedly by applying heat and hardened by cooling, and they can be easily recycled, which is why they are the most popular in recent commercial use, whereas thermosets provide better realization of fiber properties. Petroleum-based matrices are supplemented with bio-derived matrices, and PLA, which has higher strength and stiffness with natural fiber than PP, is used in terms of mechanical properties (Faruk et al., 2012).

1.4.3 Interface Strength

The mechanical properties of composites are largely determined by the interface bonding between fiber and matrix. Stress is transferred b/w the matrix and the fiber; good interface bonding is needed for optimum reinforcement; a strong interface can allow crack propagation, reducing toughness and strength. Plant-based fiber composites have limited interaction b/w hydrophilic fibers and hydrophobic matrices, resulting in poor interfacial bonding, mechanical performance limitations, and reduced moisture resistance, which impacts long-term properties. For bonding, fiber and matrix are brought into close proximity; wettability is a prerequisite. Interfacial defects caused by insufficient fiber wetting act as tress concentrators (Chen et al., 2006). The strength, toughness, tensile, and flexural strength of composites are all affected by fiber wettability (Wu & Dzenis, 2006). Chemical and physical treatments can improve the fiber's wettability as well as its interfacial strength (Bénard et al., 2007). Electrostatic bonding, mechanical interlocking, inter-diffusion bonding, and chemical bonding are examples of interfacial bonding mechanisms (Ragoubi et al., 2010). When the fiber surface is rough, mechanical interlocking occurs, which increases interfacial shear strength but has little effect on transverse tensile strength. Only metallic surfaces are suitable for electrostatic bonding. When chemical groups on the fiber surface and in the matrix react to form bonds, this is known as chemical bonding. Chemical bonding is achieved by using a coupling agent as a link between the fiber and the matrix. Inter-diffusion bonding occurs when the atoms and molecules of the fiber and matrix interact at the interface. Physical and chemical approaches are being investigated in order to improve bonding in NFCs. Corona, ultraviolet (UV), plasma, heat treatments, and electron radiation are examples of physical approaches. Plasma is generated by applying high voltage to sharp electrode tips separated by quartz at low temperature and atmospheric pressure, which is used in corona treatment. It causes chemical and physical changes in fibers, such as increased surface polarity and roughness, but it's difficult to apply to three-dimensional fibers. Plasma treatment is similar to corona treatment, but it is done in a vacuum chamber with a constant supply of gas at the right pressure and composition. Plasma treatment increases fiber surface roughness and hydrophobicity, which improves interfacial adhesion (Sinha & Panigrahi, 2009).

1.4.4 FIBER DISPERSION

Fiber dispersion is a significant element determining the characteristics of short fiber composites, and it is especially difficult for NFCs, which typically have hydrophilic fibers and hydrophobic matrix. Use of longer fiberscan increases their tendency to agglomerate. Good fiber dispersion promotes good interfacial bonding, reducing voids by ensuring that the fibers are fully surrounded by the matrix (Heidi et al., 2011). Dispersion can be influenced by processing parameters such as temperature and pressure; additives such as stearic acid to increase interfacial bonding which increase the fiber matrix interaction. The use of intensive process mixing process such as twin-screw extruder rather than a single-screw extruder leads to better fiber dispersion; this is generally at the cost of fiber damage, and fiber lengths are found to reduce dramatically during the process due to temperature and screw configuration (Beckermann & Pickering, 2008).

1.4.5 FIBER ORIENTATION

When the fiber is aligned parallel to the direction of the applied load, composites have the finest mechanical properties. Alignment is more difficult with natural fibers than with continuous synthetic fibers (Joseph et al., 1999). Prior to matrix impregnation, natural fibers are carded and placed in sheets to achieve a higher degree of fiber alignment. Traditional textile processing techniques, such as spinning, can be used to produce a continuous yarn. Wrap spinning, a method used in the textile industry since the 1970s, produces aligned fiber yarns. During compression molding, the continuous strand can be made from the same type of fiber as the short strand or from the matrix material. Thermoplastic fiber that is going to be converted into a matrix can also be used to align in the yarn direction and act as a support for the natural fiber (Vinay et al., 2014). Continuous fiber tape was created by using the fibers' own pectin as an adhesive, which was applied with a water mist and then dried while stretched. When it comes to the degree of influence of orientation on NFC mechanical performance, increasing the fiber orientation angle relative to the test direction results in a large reduction in strength and young's modulus (Raja et al., 2017).

1.4.6 POROSITY

Porosity has a significant impact on composite mechanical properties, and much effort has been put into reducing it in synthetic fiber composites. It occurs as a result of the presence of air during processing, the limited wettability of fibers, lumens, and other hollow features within the fiber bundles, and low ability of fibers to compact (Madsen et al., 2009). Porosity in NFCs increases with fiber content, increasing more quickly once the geometrical compaction limit has been exceeded, and is dependent on fiber type and orientation; flax/PP composites had porosity increase from 56 to 72 m% (Madsen & Lilholt, 2003). Its inclusion in models improves strength and stiffness prediction (Raja et al., 2017).

1.5 TYPES OF MECHANICAL TESTING IN NATURAL FIBERS

Mechanical properties of natural fiber, like metals, must be tested and improved in order to be used as green composites. Tensile strength, flexural strength, impacts, fatigue, and creep are the most important properties here. Natural fibers are commonly used to reinforce polymers because of their high strength, stiffness, and low density (Al-oqla et al., n.d.).

1.5.1 TENSILE TESTING OF BIO-COMPOSITES

In bio-composites, as in metals, the tensile test is the most important. The properties derived from the tensile test are critical in determining the best cellulosic fiber for a specific application. Furthermore, tensile properties can be improved; for example, to improve the tensile strength of HDPE/hemp fiber composites, silane and matrix-resin pre-impregnation of the fiber should be performed. Experiments revealed that the longitudinal tensile strength of silane-treated fibers increased dramatically. Transverse tensile strength, on the other hand, was not significant in natural composites. Due to the longitudinal fiber direction is much greater than the transverse fiber direction, the performance of natural composites is clearly fiber controlled (Kumar Asheesh et al., 2017). The following are some general experimental conclusions in the field of natural fiber composites:

- Due to inadequate adhesion b/w the matrix and the fibers, the tensile strength of specific natural fiber/polymer composites tends to decrease with fiber loading.
- Chemically treated natural fibers may have a higher tensile strength.
- Moisture absorption, the effect of surface treatment (using NaOH), the performance of hybrid natural fiber reinforced polymer composites, fiber size, and fiber orientations in the composites should all be considered when studying the tensile properties of thermoplastic matrices.
- The following factors should be considered when determining the tensile properties of natural fiber reinforced thermosets using siloxane treatment of polyester-based and epoxy-based composites: the effect of moisture absorption, the effect of fiber volume fraction, and the impact of fiber orientation in composites, as well as the temperature and the effects of a different geometry in the composites.

(Faruk et al., 2014; Anshuman et al., 2014;
Fávaro, 2010; George, 2013)

1.5.2 IMPACT TEST OF BIO-COMPOSITES

The impact strength of bio-fiber reinforced composites is a significant challenge. To overcome this challenge, new fiber manufacturing techniques must be continuously improved, as well as of filler/matrix adhesion. The Charpy impact test revealed increases in impact strength, which in some cases was doubled, compared to published values. The impact strength of four types of boards made from HDPE and rice straw components (rice husk, straw leaf, straw stem, and whole rice straw) formed by melting and compression molding was compared (Shah & Lakkad, 1981). The results showed that panels with rice husk had the best impact strength. The impact properties of boards made from leaf, stem, and whole straw fibers, on the other hand, did not change significantly. Recycled HDPE composites, on the other hand, outperformed virgin HDPE composites in terms of impact strength (Shah & Lakkad, 1981). Furthermore, they have shown that the toughness of natural composites can be affected by a variety of factors such as matrix intrinsic properties, fiber volume fraction, and filler-matrix bond strength (Anyakora, 2013).

1.5.3 MANUFACTURING PROCESSES OF BIO-COMPOSITES

The joining of a resin, a curing agent, some type of filler, and sometimes a solvent is a general attribute in all polymeric-based composite methods. In most cases, heat and pressure are used to shape and cure the mixture into a finished product. The resin in natural fiber reinforced composites materials holds the fillers together, protects them, and transfers the load to the fibers in the composite. On the other hand, the curing agent (or hardener) usually acts as a catalyst to speed up the curing of the resin into a hard plastic. Several green composite manufacturing processes are commonly used, including:

1.5.4 HAND LAY-UP

After being designed to the desired shape, the fibers are trimmed and spread over a mold. It's possible that some layers are required. After that, a vacuum bag is wrapped around the lay-up, where vacuum is used to remove air, compress the part, and create a barrier for the assembly when it is placed in an autoclave for curing under both heat and pressure.

1.5.5 RESIN TRANSFER MOLDING

Resin transfer molding is used to create parts with a smooth surface and low pressure. Fibers are typically laid by hand into a mold, which is then filled with a resin mixture. The part is then heat and pressure cured. There are several advantages to resin transfer molding, including

1. Big and complex shapes can be made quickly and affordably.
2. It is faster than the lay-up process.
3. Low clamping pressures is required.
4. Surface definition is better than lay-up.
5. Special fillers and inserts can be easily added.
6. Unskilled operators are possible.
7. Consistency in part is a good thing.
8. Toxic chemical exposure to workers is kept to a minimum.

1.5.6 PULTRUSION

Pultrusion is a process that involves pulling continuous roving strands from a creel into a resin bath via a strand-tensioning device. The coated strands are then cured by passing them through a heated die (Kim, 2011). After that, the continuous cured part is cut to the desired lengths. It has a number of advantages (Kim, 2011) including: 1. Effective material utilization. 2. High throughput is possible. 3. Resin content is good.

The following are the disadvantages: 1. Requires a uniform cross section. 2. Fiber and resin may build up at the die opening, causing jamming and breakage due to increased friction. 3. If too much resin is used, the part's strength will suffer. 4. If the die does not conform well to the fibers being pulled, voids can occur. 5. The quick curing system may reduce the strength of the manufactured part.

1.5.7   EXTRUSION PROCESSES

It involves forcing a thick, viscous liquid through small holes in a spinneret to form continuous semi-solid polymer strings (filaments). In the case of thermoplastic synthetic polymers, this process is usually achieved by heat and pressure; non-thermoplastic polymers can be processed by dissolving in a suitable solvent or by heat. New technologies for some specialty fibers made of polymers that do not melt, dissolve, or form appropriate derivatives when reacting with small fluid molecules have recently been developed. The spinneret is the most important part of this process. The spinneret resembles a bathroom shower head, in that it is a corrosion-resistant metal plate with a slew of small holes. Because those holes are so small, they require special attention, such as filtering the liquid that will feed them, regular maintenance, and a schedule for disassembling and cleaning them. Wet, dry, melt, and gel spinning are the four methods for spinning filaments (solidification of the liquid polymer when exiting the spinneret) of manufactured fibers.

1.5.8   WET SPINNING

Fiber-forming substances that have been dissolved in a solvent are given this name. The spinnerets are immersed in a chemical bath, and as the filaments emerge from the solution, they precipitate and solidify. Wet spinning refers to the process of making fibers in which the solution is extruded directly into the precipitating liquid. This method can be used to make acrylic, rayon, aramid, modacrylic, and spandex.

1.5.9   DRY SPINNING

Solidification is achieved by evaporating the solvent in a stream of air or inert gas, similar to wet spinning but without the use of a chemical reaction.

1.5.10   MELT SPINNING

The fiber-forming substance is melted for extrusion through the spinneret and then solidified by cooling in melt spinning. This method can produce a variety of cross-sectional shapes, including round, trilobal, pentagonal, octagonal, and others. For example, using pentagonal-shaped and hollow fibers in carpet construction is recommended to reduce soil and dirt.

1.5.11   GEL SPINNING

It's used to make special fiber properties like high strength. By keeping the polymer in a thick liquid state during the extrusion process, the tensile strength can be increased. In liquid crystal form, the polymer chains will bind together at various points.

1.5.12 NATURAL FIBER INJECTION PROCESS

For its simplicity and quick processing cycle, the natural fiber injection process is the most widely used molding method in the industry for producing polymer composites. A pre-calculated amount of matrix and fiber mixture is mixed and injected into the mold by an injection molding machine, resulting in the desired product. The injection unit, mold, and ejection and clamping unit are the three major sections. A heated screw barrel with a compression screw is included in the injection unit.

Each part's purpose is obvious from the name. Specifically, the heated barrel heats the polymer matrix to melt before injection, the screw transports and compresses pellets from the hopper into the heated barrel, mixes the polymer matrix and fiber, and injects the mixture into the closed mold with the final shape of the product. The molds are identical to any other mold used in the injection process. Molds are made using Computer Numerical Control (CNC) machining processes, and they have cooling/heating coils to regulate the mold's temperature, as well as ejectors to eject the finished product once the processes are finished.

1.6 APPLICATIONS

Fibers from nature reinforced composites are rapidly becoming a viable alternative to metal or ceramic-based materials in a variety of industries, including automotive, marine, aerospace, sporting goods, and electronics. Natural fibers have good specific properties, but there is a lot of variation in them. Within the United States, straw is being used as a composite material for construction. Straw is used in the construction of buildings. Natural composites, mostly based on polyester or polypropylene and fibers like hemp, flax, jute, or sisal, are already used in many automotive components. Rather than technical demands, the use of natural fiber in the industry is driven by price, light weight, and marketing.

Natural fiber composites are widely used in Germany. Natural composites are used in interior and exterior applications by German automakers such as Audi, Mercedes Benz, and Volkswagen. Jute-based composites would be used for structural applications, such as housing interior elements. Mirror casing, projector cover, helmet, voltage stabilizer, and roof have all been made from coir/polyester composites. NFCs are also being used in the aircraft industry for interior paneling. They've been used as a substitute for synthetic fiber in a variety of applications, including electronic devices, packaging, marine railings, and sporting goods. However, some disadvantages exist, such as moisture adsorption, limited processing temperature, and variable quality, which limit their performance (Ragoubi et al., 2010; Ayrilmis et al., 2011).

1.7 METHODS FOR IMPROVING MECHANICAL CHARACTERISTICS

1.7.1 CHEMICAL TREATMENT

The hydrophilic nature of natural fibers makes them incompatible with the hydrophobic polymer matrix. The hydrophilic nature of cellulosic fiber has an impact on the fiber–matrix interface. Grafting of monomers, bleaching, acetylation, and other treatments are used for this purpose. The thermal stability of natural fibers can also be improved by grafting fibers with monomers. Natural fibers are pre-treated to help chemically modify or clean the fiber surface (Edeerozey et al., 2007). Since cellulose and lignin contain hydroxyl groups, natural fibers are amenable to modification. This hydroxyl group participates in hydrogen bonding within the cellulose, lowering activity towards the matrix. As a result, a variety of treatments are used to improve the strength, ageing, and fiber matrix adhesion of natural fiber composites (Ali, 2018).

1.7.2 Alkaline Treatment

Natural fiber's hydrophilic nature necessitates chemical modification to improve interfacial properties b/w fiber and resin (Li et al., 2007). Alkali treatment is one of the most cost-effective and environmentally friendly chemical modification methods. By removing cellulosic content that covers the external surface of the fiber cell wall, alkaline treatment increases surface roughness and the amount of exposed cellulose on the fiber surface, resulting in better mechanical interlocking. Numerous alkali treatment trials for natural fibers have been documented in previous studies. (Atiqah et al., 2014) treated kenaf fiber with a 6% sodium hydroxide (NaOH) solution for three hours and obtained the satisfactory performance in terms of flexural, tensile, and impact strength. (Merlini et al., 2011) tried an hour of alkaline treatment on banana short fibers with a 10% NaOH solution. . . . Previous work by (Asim, 2016) previous research on alkali treatment of pineapple fiber revealed that fibers treated with 6% NaOH had improved mechanical properties. When compared to untreated fibers, the alkaline treatments showed improved mechanical properties. The goal of this research was to find the best alkaline treatment for single pineapple leaf fiber strands and see how it affected mechanical properties, surface micrography, heat resistivity, and interfacial bonding with epoxy matrix (Zin et al., 2018).

1.7.3 Sodium Chlorite Treatment

(Arifuzzaman Khan, 2009) chemically modified the okra bast fiber with $NaClO_2$ to improve mechanical and thermal properties as well as increase hydrophobicity. The Fourier transform infrared (FTIR) spectroscopy was used to determine the extent of the modification reaction. The morphology and crystalline index of jute fibers were studied using scanning electron microscopy and wide angle X-ray diffractometry. Chemical treatment of the fiber surface resulted in a noticeable difference in its tensile properties (i.e. tensile strength, young's modulus, and extension at break). Based on the activation of sodium chlorite by hexamethylenetetramine (HMTA) in the presence of a nonionic wetting agent, (Zahran et al., 2005) developed a novel chemical formulation for bleaching flax fibers in one step. They claim that when HMTA is properly formulated, it activates the decomposition of $NaClO_2$ to liberate nascent oxygen rather than chlorine dioxide (Ali et al., 2018).

1.7.4 Treatment With Methacrylate

Flax fibers were esterified to make them hydrophobic, according to (Cantero et al., 2003). They made a flax/PP composite by soaking it in 10% methacrylate (MA) for

25 hours at 50°C. The flax fiber composites had good flexural and tensile strength, according to the researchers. (Kaith & Kalia, 2007) used 20wt% methyl methacrylate (MMA) to treat flax fibers for 20 minutes before using phenolic resin to create a flax composite. They noticed that the treated flax fiber composites absorbed less moisture (John & Anandjiwala, 2008).

1.7.5 SILANE TREATMENT

Silanes act as coupling agents and stabilize composite materials by allowing glass fibers to adhere to a polymeric matrix. Fiber hydroxyl groups are reduced by silane coupling agents, resulting in a better interface. In the presence of moisture, a hydrolyzable alkoxy group leads to the formation of silanols. The fibers' hydroxyl groups react with silanol to form stable covalent bonds with the cell wall, which are chemisorbed on the fiber surface. As a result of the covalent bonding, silane prevents the fibers from swelling by forming a cross-linked network. The flax fibers were treated with silane to make them hydrophobic, and composites were made with both treated and untreated flax fibers (Alix et al., 2011). The interface of the composite was studied using UP as a matrix. The use of silane coupling agents can improve the degree of crosslinking in the interface region. Silane coupling agents are ideal for modifying the interfacial structure of natural fiber matrixes. The sol gel process is best for improving interfacial adhesion and improving the mechanical properties of fiber/polymer composites. Silane is absorbed and condensed on the surface of the fiber. The hydroxyl group of natural fibers and the silane coupling agent form hydrogen bonds. By heating the treated fibers to high temperatures, this link can be converted to covalent bonds (Rana et al., 2021; Xie et al., 2010).

1.7.6 ACETYLATION

Acetylation is a well-known esterification method. Acetylation of natural fibers causes plasticization in natural cellulosic fibers, which is commonly used to protect the cell wall of wood cellulosic fibers from moisture and temperature changes, as well as to improve dimensional stability. When acetic anhydride is applied to lignocellulose, it reacts with the hydroxyl groups of the cellulose and also prevents the diffusion of the reagents, as (Yao et al., 2008) noticed after treating flax fiber with acetic anhydride. They found an 18% increase in degree of acetylation, as well as significant increases in tensile and flexural strength. An acetyl functional group (CH3COO–) is introduced into the fiber structure during the acetylation reaction. Acetic acid (CH3COOH) is produced as a by-product of the reaction, which must be removed from the lignocellulosic material before the fiber can be used. The acetic anhydride (CH_3–C(1O)–O–C(1O)–CH_3) chemical modification replaces the polymer hydroxyl groups of the cell wall with acetyl groups, changing the properties of these polymers to make them hydrophobic (Li et al., 2007). Anhydride treatment is usually done with PP or polyethylene (PE) in a toluene or xylene solution. It is impregnated with fibers to carry out the hydroxyl group reaction on the fiber surface. (Hughes et al., 2007) noticed that treating the fiber surface with propionic anhydride (PA) and methacrylic showed better yield properties. (Alatriste-Mondragon et al., 2003) described two types of modification mechanisms:

(a) introducing a reactive vinylic group via methacrylic anhydride (MA) and (b) coating hydrocarbons on the surface via polyacrylic anhydride (PA). They tested the strength of a flax fiber composite made of methacrylic polypropylene (MAPP). The strength of the MA-treated flax fiber composite was lower than that of the unmodified flax fiber composite.

1.7.7 MERCERIZATION

It is a process in which alkali is used to treat natural fibers, causing fibrillation and the breakdown of fiber bundles into smaller fibers. As a result, a rough surface topography develops, resulting in improved fiber matrix interface adhesion and improved mechanical properties (Joseph et al., 2002). Furthermore, the mercerization process increases fiber active sites, allowing for better fiber wetting. Chemical composition, polymerization degree, and molecular orientation are all influenced, and these factors have a long-term impact on mechanical properties. The alkali treatment also aids in the improvement of cotton fiber composite properties (Koyuncu et al., 2016). When treating flax fiber with 2.5%, 5%, 10%, 15%, 18%, 20%, 25%, and 30% sodium hydroxide, (Sreekala et al., 2000) found that a 10–30% solution gave the best results. They discovered that the best concentrations for mercerization were 5%, 18%, and 10%. Several researchers carried out mercerization and found that it increased the amount of amorphous cellulose while also removing hydrogen bonding (Misra et al., 2002). While removing the lignin and hemicellulose, the mercerization process affects the chemical composition of natural fiber, as well as the degree of polymerization and molecular orientation of cellulose crystallites. The intensity ratio of the stretching modes of symmetric (C–O–C) and asymmetric proposed by (Bledzki et al., 2004) could be seen by FT Raman spectroscope. The chemical properties of mercerized flax fiber composite were investigated by (Wang et al., 2007). Polystyrene was used as the matrix and the treated flax fibers were used as reinforcement. Mercerization of flax fiber improved the mechanical properties of polystyrene composites, they noticed. Similarly, (Bledzki et al., 2008) investigated the effect of acetylation on flax/PP composites and discovered that the compositional change caused by the removal of lignin and hemicellulose resulted in an increase in mechanical properties as well as an increase in thermal stability of treated flax fiber. (Yan et al., 2012) used alkali treatment of flax fiber to improve the mechanical properties of natural fiber composites. They made the flax epoxy composite after using 5% NaOH for 30 minutes. They saw a significant increase in tensile strength (21.9%) and flexural strength (16%), as well as a slight increase in transverse strength. When compared to untreated natural fiber based composites, the pretreated natural fiber based composites performed better in terms of mechanical properties (Patel et al., 2018; Ali et al., 2018).

1.7.8 ETHERIFICATION

Natural fiber composites can be etherified to make them more useful and improve certain properties (Kalia et al., 2009). The charged intermediate species formed by sodium hydroxide and the fiber allows for faster nucleophilic addition of epoxides, alkyl halides, benzyl chloride, acrylonitrile, and formaldehyde (Kalia et al., 2009). At

40°C, the cellulosic materials are reacted with acrylonitrile in a 4% NaOH aqueous solution saturated with NaSCN as a swelling agent and catalyst. Ethylified natural fiber can be made by reacting benzyl chloride with hydroxyl group (Mohanty et al., 2001). The results showed that treated fiber had the same structure and morphology as the untreated fiber, but it had the chemical groups on its surface that were needed for reinforcing applications in polymer composites (Ali et al., 2018).

1.7.9 ENZYMATIC TREATMENT

When enzymes are used in conjunction with chemical and mechanical methods for material modification, enzymatic treatment is a very relevant & helpful step. Enzymes are effective catalysts that work in a highly specific manner under low-energy conditions. Oxidative enzymes, such as peroxidases, can be used to further functionalize lignocelluloses. Laccase converts phenolic hydroxyls to phenolic radicals in the presence of oxygen. It was also discovered that laccase treatment reduced the lignin content of single cellulosic fibers from 35% to 24% (Kudanga et al., 2011). Laccase, when combined with natural phenols like acetosyringore, P-coumaric acid, and syringe-aldehyde, gave natural fiber composites antimicrobial properties (flax composite) (Ali et al., 2018).

1.7.10 TREATMENT WITH ISOCYANATE

The isocyanate functional group (–N1C14O) in an isocyanate compound is highly susceptible to reaction with the hydroxyl groups of cellulose and lignin in fibers. In fiber-reinforced composites, isocyanate is said to act as a coupling agent. Dried alkali-treated fibers were soaked in an appropriate volume of carbon tetrachloride (CC14) and a small (1 ml) dibutyl tin dilaurate catalyst in a round bottomed flask. After that, a pressure equalizing funnel containing the urethane derivative was fitted to a round bottomed flask, and the urethane derivative was added dropwise into the flask with constant stirring. After the addition of urethane, the reaction was allowed to continue for another hour. These urethane-treated fibers were purified with acetone by refluxing for eight hours in a Soxhlet apparatus, then washing with distilled water and drying in an oven at 80°C (Joseph et al., 1996).

1.7.11 PEROXIDE TREATMENTS

Peroxide treatments have piqued the interest of most researchers for the treatment of cellulosic fibers because they are simple and provide good mechanical properties. As proposed by (Sreekala et al., 2000) organic peroxides are easily decomposed to free radicals, which then react with the cellulose of the fiber and the hydrogen group of the matrix. Fibers are treated with 6% dicummyl peroxide or benzoyl peroxideinacetone solution for 30 minutes after being pretreated with alkali. Dicumyl peroxide from an acetone solution was used to treat flax fibers. The fibers were soaked in the solution for 30 minutes at 70°C. After that, the fibers were washed in distilled water and placed in an oven at 80°C for 24 hours to improve their hydrophobic properties.

1.7.12 Benzoylation

Benzoylation is a crucial step in organic synthesis, and benzoyl chloride is the most common benzoyl derivative used in fiber treatment. Benzoyl chloride contains benzoyl (C6H5C14O), which is thought to reduce the hydrophilic nature of treated fibers and improve interaction with hydrophobic matrixes. For the surface treatment of sisal fibers (Joseph et al., 2002) used a solution of sodium hydroxide and benzoyl chlorite (C6H5COCl). To activate the hydroxyl group of lignin and cellulose in fibers, they were alkaline pre-treated. The fibers were then immersed for 15 minutes in a 10% NaOH and benzoyl chloride solution. The fibers were then washed and dried at 80°C for 24 hours after being soaked in ethanol for one hour to remove benzoyl chloride. After the treatment, the surface was modified and the hydrophobicity improved.

1.7.13 Plasma Treatment

This is a highly successful treatment approach for modifying the surface characteristics of natural polymers while leaving the bulk properties unchanged. To generate the plasma discharge, either cold plasma therapy or corona treatment can be utilized. Both are plasma treatment technologies in which an ionized gas comprises an equal number of positively and negatively charged molecules that react with the surface of the material. The frequency of the electric discharge is the primary distinction b/w the two forms of plasmas. Microwave energy can create cold plasma with a high frequency, whereas corona plasma is created by an alternating current discharge with a lower frequency at atmospheric pressure. The surface modification of wood and synthetic polymer surfaces was influenced by the type of ionized gas used. (Maldas et al., 1989) described a method for activating a wood surface to improve polyolefin adhesion by exposing it to plasmas. Corona treatment or cold plasma treatment was used to create plasma discharges. Furthermore, researchers have recently investigated the number of the polar component of surface energy of pine wood for plasma modification, which includes power, sample distance from plasma source, treatment time, plasma treatment stability, and gas type. The pulp sheets with moisture contents of up to 85% were treated with a Corona discharge. The process was carried out in the presence of air and nitrogen atmospheres, and the sheets' chemical modification was evaluated using dye (Sakata et al., 1991).

1.7.14 Ozone Treatments

Exposure to ozone or oxygen-fluorine gas can alter the surface of cellulosic materials. Ozone gas was used to expose cellulose fibers, PE membranes, and films by (Hedenberg & Gatenholm, 1996). When low density polyethylene (LDPE) was treated with ozone, the adhesion properties of the composites improved. The increase in bonding strength was proposed using two mechanisms: hydrogen bonding of LDPE with carbonyl groups as revealed by spectroscopy and covalent bonding initiated by the decomposition of hydro peroxides. The ozone treatment changes the contact angles of various liquids with surfaces and raises total surface energies. Furthermore, the oxidation procedure raises the polar component of LDPE's surface energy. This increased polarity of LDPE will provide a much better foundation for interactions with the numerous hydroxyl groups on cellulose, lowering the material's hydrophilic character. By exposing PE to oxygen-fluorine gas, researchers were able to improve the strength properties of PE and pulp composites. The authors discovered that exposure increased the specific (acid–base) interaction parameter, and that hydrogen bonding b/w fluorine and hydroxyl groups, carbonyl, and hydroxyl groups were responsible for the improved properties (Ali et al., 2018).

1.7.15 GRAFTING

Vinyl monomer graft copolymerization onto wood and cellulose-based materials was first reported in 1953. Grafting has gotten a lot of attention since then for cellulose modification. When the surfaces of fibers, wood, and plant-based materials are activated, the efficiency of grafting monomers onto natural surfaces increases. (Hill & Cetin, 2000) noticed that methyl acrylic anhydride-treated wood allowed for the grafting of MMA & styrene. (Grelier et al., 1997) used microwave activation to graft isocyanate with a UV absorbing chromophore to medium density fiber boards. In comparison to the number of studies on graft copolymerization onto cellulose, research on copolymerization onto more complex materials such as wood and plant fiber is limited. Through resonance stabilization of free radicals, lignin and other extractives found in wood act as antioxidants, inhibiting polymerization and grafting. The copolymerization process is actually inhibited by lignin. Graft copolymerization is inhibited by the type of catalyst used and the nature of lignin. However, depending on the reaction conditions, lignin may become more modified than holocellulose or reduce monomer homopolymerization due to a chain transfer mechanism. The involvement of lignin in the graft copolymerization of polymethyl methacrylate onto bagasse pith initiated by Fe2 and hydrogen peroxide and potassium permanganate was described by Zheng et al. (KMNO4). Lignin was more susceptible to grafting, and the copolymerization was aided by the hydroxyl cyclohexadienyl radicals formed by lignin. In the graft copolymerization of stone ground wood pulp, the same initiators were used as in the bagasse treatment. Copolymerization was initiated by phenolic hydroxy radicals produced by reaction with ferric ions. The initial stage of delignification, which saw a 0.6% loss of lignin due to the sodium chlorite method of pulping, improved grafting efficiency, but delignification did not have the same effect. The authors also discovered that the phenyl ring formed phenoxy radicals, which either initiated copolymerization or decomposed the lignin phenyl rings, depending on reaction conditions. When KMnO4 was used as an initiator, Marchetti et al. reported the deposition of manganese dioxide (MnO2) on the wood cell wall, which contributed to the radical chain transfer onto wood. The effect of lignin on copolymerization in partially delignified jute fibers was discovered by Ghosh et al. Initial delignification was only slightly improved by sodium chlorite grafting, but grafting efficiency was doubled in the presence of only 0.8% lignin. Sugarcane fibers, both bleached and unbleached, were grafted with hydroxy phenyl benzotriazole UV absorber and piperidinyloxynitroxide radical, then treated with acetic anhydride (Ali et al., 2018).

1.7.16 FLUOROCARBON TREATMENT

For the treatment of jute fibers (Ali et al., 2018) used fluorocarbons, hydrocarbons, and hybrid fluorocarbons. The surface free energy of these chemicals is lower. It is a well-known fact that the lower a material's surface free energy, the lower the moisture regain. At a concentration of 40 g/l, a significant difference in moisture regain values was observed between treated and untreated reinforcement samples. The treated reinforcement composite had a very low moisture content and better mechanical properties (tensile and flexural strength). Because of the hybrid fluorocarbon's dual nature, the treated jute fibers and associated composites demonstrated superior qualities than the other two chemicals (hydrophilic and hydrophobic groups).

1.8   CONCLUSION

The mechanical performance of bio-composites is governed by the interfacial bonding between their constituents. To reduce failure inside composites, the most constructive reinforcement conditions must be achieved. The fabrication process would have a significant impact on the final behavior of the green composites because it could destroy the reinforcement's desired properties. As a result, failure stresses of materials must be assessed for design purposes. Experimentation and/or Finite Elements Analysis (FEA) techniques can be used to accomplish this. Failure theories developed for metals or other isotropic materials do not apply to green composites because they are neither isotropic nor show gross yielding. Instead, many new failure theories for bio-composites have been proposed. Future opportunities for eco-composite research can be divided into two categories. One is being carried out to find new ways to make low-cost biodegradable polymers with better mechanical and thermal properties. The second area of research is a "cost effective" modification of NFs, as the competitive cost of NFs is the main market attraction of eco-composites. The reports comparing the LCA assessment of NF composites with glass fiber reinforced composites found that NF composites are superior in specific automotive industry applications, first and foremost due to their lower weight. Further evaluation of the eco performance of NF composites, as well as LCA studies, should confirm their "green" status (Avella et al., 2007).

REFERENCES

Alatriste-Mondragon, F., Iranpour, R., & Ahring, B. K. (2003). Toxicity of di-(2-ethylhexyl) phthalate on the anaerobic digestion of wastewater sludge. *Water Research*, *37*(6), 1260–1269.

Ali, A., Shaker, K., Nawab, Y., Jabbar, M., Hussain, T., Militky, J., & Baheti, V. (2018). Hydrophobic treatment of natural fibers and their composites—A review. *Journal of Industrial Textiles*, *47*(8), 2153–2183. https://doi.org/10.1177/1528083716654468

Al-oqla, F. M., Almagableh, A., & Omari, M. A. (n.d.). Design and fabrication of green biocomposites Á design of composites Á biocomposites Á. In *Green Biocomposites* (pp. 45–67). https://doi.org/10.1007/978-3-319-49382-4

Al-oqla, F. M., Salit, M. S., & Ishak, R. (2014). Combined multi-criteria evaluation stage technique as an agro waste evaluation indicator for polymeric composites: Date palm fibers as a case study. *BioResources*, *9*(3), 4608–4621.

Al-oqla, F. M., & Sapuan, S. M. (2014). Natural fi ber reinforced polymer composites in industrial applications: Feasibility of date palm fi bers for sustainable automotive industry. *Journal of Cleaner Production*, *66*, 347–354. https://doi.org/10.1016/j.jclepro.2013.10.050

Alix, S., Lebrun, L., Morvan, C., & Marais, S. (2011). Study of water behaviour of chemically treated flax fibers-based composites: A way to approach the hydric interface. *Composites Science and Technology*, *71*(6), 893–899. https://doi.org/10.1016/j.compscitech.2011.02.004

Anshuman, S., & Madhusoodan, M. (2014). Preparation and mechanical characterization of epoxy based composites developed by bio waste. *International Journal of Research in Engineering and Technology*, *4*(4).

Anyakora, A. N. (2013, May). *Investigation of impact strength properties of oil and date palm frond fiber reinforced polyester composites.* https://inpressco.com/investigation-of-impact-strength-properties-of-oil-and-date-palm-frond-fiber-reinforced-polyester-composites/

Aparecido, P., & Giriolli, J. C. (n.d.). *Natural fibers plastic composites for automotive applications characteristics of natural fibers*, pp. 1–9.

Arifuzzaman Khan, G. M., Shaheruzzaman, M., Rahman, M. H., Abdur Razzaque, S. M., Islam, M. S., & Alam, M. S. (2009). Surface modification of okra bast fiber and its physico-chemical characteristics. *Fibers and Polymers*, *10*(1), 65–70. https://doi.org/10.1007/s12221-009-0065-1

Asim, M., Jawaid, M., Abdan, K., & Ishak, M. R. (2016). Effect of alkali and silane treatments on mechanical and fiber-matrix bond strength of kenaf and pineapple leaf fibers. *Journal of Bionic Engineering*, *13*(3), 426–435. https://doi.org/10.1016/S1672-6529(16)60315-3

Atiqah, A., Maleque, M. A., Jawaid, M., & Iqbal, M. (2014). Development of kenaf-glass reinforced unsaturated polyester hybrid composite for structural applications. *Composites Part B: Engineering*, *56*, 68–73. https://doi.org/10.1016/j.compositesb.2013.08.019

Avella, M., Malinconico, M., Buzarovska, A., Grozdanov, A., Gentile, G., & Errico, M. E. (2007). Natural fiber eco-composites. *Polymer Composites*, *28*, 98–107. https://doi.org/10.1002/pc.20270

Ayrilmis, N., Jarusombuti, S., Fueangvivat, V., & Bauchongkol, P. (2011). Coir fiber reinforced polypropylene composite panel for automotive interior applications. *Fibers and Polymers*, *12*(7), 919–926. https://doi.org/10.1007/s12221-011-0919-1

Beckermann, G. W., & Pickering, K. L. (2008). Engineering and evaluation of hemp fiber reinforced polypropylene composites: Fiber treatment and matrix modification. *Composites Part A: Applied Science and Manufacturing*, *39*(6), 979–988. https://doi.org/10.1016/j.compositesa.2008.03.010

Beg, M. D. H. (2007). *The improvement of interfacial bonding, weathering and recycling of wood fiber reinforced polypropylene composites* [Doctor Dissertation], University of Waikato, Waikato. https://hdl.handle.net/10289/2553

Bénard, Q., Fois, M., & Grisel, M. (2007). Roughness and fiber reinforcement effect onto wettability of composite surfaces. *Applied Surface Science*, *253*, 4753–4758. https://doi.org/10.1016/j.apsusc.2006.10.049

Bledzki, A. K., Fink, H. P., & Specht, K. (2004). Unidirectional hemp and flax EP- and PP-composites: Influence of defined fiber treatments. *Journal of Applied Polymer Science*, *93*(5), 2150–2156. https://doi.org/10.1002/app.20712

Bledzki, A. K., Mamun, A. A., Lucka-Gabor, M., & Gutowski, V. S. (2008). The effects of acetylation on properties of flax fiber and its polypropylene composites. *Express Polymer Letters*, *2*(6), 413–422. https://doi.org/10.3144/expresspolymlett.2008.50

Cantero, G., Arbelaiz, A., Llano-Ponte, R., & Mondragon, I. (2003). Effects of fibre treatment on wettability and mechanical behaviour of flax/polypropylene composites. *Composites Science and Technology*, *63*(9), 1247–1254.

Charlet, K. (2007). Characteristics of Hermès flax fibres as a function of their location in the stem and properties of the derived unidirectional composites. *Composites Part A: Applied Science and Manufacturing*, *38*, 1912–1921. https://doi.org/10.1016/j.compositesa.2007.03.006

Chen, P., Lu, C., Yu, Q., Gao, Y., Li, J., & Li, X. (2006). Influence of fiber wettability on the interfacial adhesion of continuous fiber-reinforced PPESK composite. *Journal of Applied Polymer Science*, *102*, 2544–2551. https://doi.org/10.1002/app.24681

Devi, L. U. M. A., Bhagawan, S. S., & Thomas, S. (1996). Mechanical properties of pineapple leaf fibre reinforced polypropylene composites. *Materials & Design*, 3–7.

Dubrovski, D. P., & Golob, D. (2009). Effects of woven fabric construction and colour on ultraviolet protection. *Textile Research Journal*, *79*(4), 351–359, ISSN 0040-5175.

Edeerozey, A. M. M., Akil, H., Azhar, A. B., & Ariffin, M. I. Z. (2007). Chemical modification of kenaf fibers. *Materials Letters*, *61*, 2023–2025. https://doi.org/10.1016/ j.matlet.2006.08.006

Fadzli, M., Abdollah, B., Fazillah, F., Ismail, N., Amiruddin, H., & Umehara, N. (2015). Selection and verification of kenaf fibers as an alternative friction material using Weighted Decision Matrix method. *Materials & Design*, *67*, 577–582. https://doi.org/10.1016/j.matdes.2014.10.091

Faruk, O., Bledzki, A. K., Fink, H., & Sain, M. (2012). Progress in polymer science biocomposites reinforced with natural fibers: 2000–2010. *Progress in Polymer Science*, *37*(11), 1552–1596. https://doi.org/10.1016/j.progpolymsci.2012.04.003

Faruk, O., Bledzki, A. K., Fink, H., & Sain, M. (2014). Progress report on natural fiber reinforced composites. *Macromolecular Materials and Engineering*, 9–26. https://doi.org/10.1002/ mame.201300008

Fávaro, S. L., Ganzerli, T. A., de Carvalho Neto, A. G. V., da Silva, O. R. R. F., & Radovanovic, E. (2010). Chemical, morphological and mechanical analysis of sisal fiber-reinforced recycled high-density polyethylene composites. *Express Polymer Letters*, *4*(8), 465–473. https://doi.org/10.3144/expresspolymlett.2010.59

George, G., Joseph, K., Nagarajan, E. R., Tomlal Jose, E., & George, K. C. (2013). Dielectric behaviour of PP/jute yarn commingled composites: Effect of fiber content, chemical treatments, temperature and moisture. *Composites Part A: Applied Science and Manufacturing*, *47*, 12–21. https://doi.org/10.1016/j.compositesa.2012.11.009

Grelier, S., et al. (1997). Attempt to protect wood colour against UV/visible light by using antioxidants bearing isocyanate groups and grafted to the material with microwave. *Holzforschung*, *51*, 511–518.

Hedenberg, P., & Gatenholm, P. (1996). Conversion of plastic/cellulose waste into composites. II. Improving adhesion between polyethylene and cellulose using ozone. *Journal of Applied Polymer Science*, *60*(13), 2377–2385.

Heidi, P., Bo, M., Roberts, J., & Kalle, N. (2011). The influence of biocomposite processing and composition on natural fiber length, dispersion and orientation. *Journal of Materials Science and Engineering*, *1*, 190–198.

Hill, C. A. S., & Cetin, N. S. (2000). Surface activation of wood for graft polymerization. *International Journal of Adhesion and Adhesives*, *20*(1), 71–76. https://doi.org/10.1016/ S0143-7496(99)00017-2

Holbery, J., & Houston, D. (2006). Natural-fiber-reinforced polymer composites in automotive applications. *JOM*, *58*(11), 80–86. https://doi.org/10.1007/s11837-006-0234-2

Hughes, M., Carpenter, J., & Hill, C. (2007). Deformation and fracture behaviour of flax fiber reinforced thermosetting polymer matrix composites. *Journal of Materials Science*, *42*(7), 2499–2511. https://doi.org/10.1007/s10853-006-1027-2

John, M. J., & Anandjiwala, R. D. (2008). Recent developments in chemical modification and characterization of natural fiber-reinforced composites. *Polymer Composites*, *29*(2), 187–207. https://doi.org/10.1002/pc.20461

Joseph, K., Thomas, S., & Pavithran, C. (1996). Effect of chemical treatment on the tensile properties of short sisal fiber-reinforced polyethylene composites. *Polymer*, *37*(23), 5139–5149. https://doi.org/10.1016/0032-3861(96)00144-9

Joseph, P. V, Joseph, K., & Thomas, S. (1999). Effect of processing variables on the mechanical properties of sisal-fiber-reinforced polypropylene composites. *Composites Science and Technology*, *59*(11), 1625–1640. https://doi.org/10.1016/S0266-3538(99)00024-X

Joseph, S., Sreekala, M. S., Oommen, Z., Koshy, P., & Thomas, S. (2002). A comparison of the mechanical properties of phenol formaldehyde composites reinforced with banana fibres and glass fibres. *Composites Science and Technology*, *62*(14), 1857–1868.

Kaith, B. S., & Kalia, S. (2007). Grafting of flax fiber (Linum usitatissimum) with vinyl monomers for enhancement of properties of flax-phenolic composites. *Polymer Journal*, *39*(12), 1319–1327. https://doi.org/10.1295/polymj.PJ2007073

Kalia, S., Kaith, B. S., & Kaur, I. (2009). Pretreatments of natural fibers and their application as reinforcing material in polymer composites—A review. *Polymer Engineering & Science*, *49*(7), 1253–1272. https://doi.org/10.1002/pen.21328

Koyuncu, M., Karahan, M., Karahan, N., Shaker, K., & Nawab, Y. (2016). Static and dynamic mechanical properties of cotton/epoxy green composites. *Fibers and Textiles in Eastern Europe*, *24*(4), 105–111. https://doi.org/10.5604/12303666.1201139

Kudanga, T., Nyanhongo, G. S., Guebitz, G. M., & Burton, S. (2011). Potential applications of laccase-mediated coupling and grafting reactions: A review. *Enzyme and Microbial Technology*, *48*(3), 195–208. https://doi.org/10.1016/j.enzmictec.2010.11.007

Kumar, A., & Srivastava, A. (2017). Preparation and mechanical properties of jute fiber reinforced epoxy composites. *Industrial Engineering & Management*, *6*, 234. https://doi.org/10.4172/2169-0316.1000234.

Le, T. M., & Pickering, K. L. (2015). The potential of harakeke fiber as reinforcement in polymer matrix composites including modelling of long harakeke fiber composite strength. *Composites Part A: Applied Science and Manufacturing*, *76*, 44–53. https://doi.org/10.1016/j.compositesa.2015.05.005

Li, X., Tabil, Æ. L. G., & Panigrahi, Æ. S. (2007). Chemical treatments of natural fiber for use in natural fiber-reinforced composites: A review. *Journal of Polymers and the Environment*, *15*, 25–33. https://doi.org/10.1007/s10924-006-0042-3

Madsen, B., & Lilholt, H. (2003). Physical and mechanical properties of unidirectional plant fibre composites—An evaluation of the influence of porosity. *Composites Science and Technology*, *63*(9), 1265–1272.

Madsen, B., Thygesen, A., & Lilholt, H. (2009). Plant fiber composites—Porosity and stiffness. *Composites Science and Technology*, *69*(7–8), 1057–1069. https://doi.org/10.1016/j.compscitech.2009.01.016

Maldas, D., Kokta, B. V., & Daneault, C. (1989). Influence of coupling agents and treatments on the mechanical properties of cellulose fiber—Polystyrene composites. *Journal of Applied Polymer Science*, *37*(3), 751–775. https://doi.org/10.1002/app.1989.070370313

Merlini, C., Soldi, V., & Barra, G. M. O. (2011). Influence of fiber surface treatment and length on physico-chemical properties of short random banana fiber-reinforced castor oil polyurethane composites. *Polymer Testing*, *30*(8), 833–840. https://doi.org/10.1016/j.polymertesting.2011.08.008

Misra, S., Misra, M., Tripathy, S. S., Nayak, S. K., & Mohanty, A. K. (2002). The influence of chemical surface modification on the performance of sisal-polyester biocomposites. *Polymer Composites*, *23*(2), 164–170. https://doi.org/10.1002/pc.10422

Mohanty, A. K., Misra, M., & Dreal, L. T. (2001). Surface modifications of natural fibers and peformance of the resulting biocomposite. *Composite Interfaces*, *8*(5), 313–343.

Panigrahi, S., Polytechnic, S., & Tabil, L. G. (2007, March). Pre-treatment of flax fibers for use in rotationally molded biocomposites pre-treatment of flax fibers for use in rotationally molded biocomposites (A thesis submitted to the college of graduate studies and research in partial fulfillment of the requirement). *Journal of Reinforced Plastics and Composites*, *26*. https://doi.org/10.1177/0731684406072526

Patel, B. C., Acharya, S. K., & Mishra, D. (2018). Environmental effect of water absorption and flexural strength of red mud filled jute fiber/polymer composite. *International Journal of Engineering, Science and Technology*, *4*(4), 49–59. https://doi.org/10.4314/ijest.v4i4.4

Pickering, K. L., Beckermann, G., Alam, S. N., & Foreman, N. J. (2007). Optimising industrial hemp fiber for composites. *Composites Part A: Applied Science and Manufacturing*, *38*(2), 461–468. https://hdl.handle.net/10289/9805

Ragoubi, M., Bienaimé, D., Molina, S., George, B., & Merlin, A. (2010). Impact of corona treated hemp fibers onto mechanical properties of polypropylene composites made thereof. *Industrial Crops and Products*, *31*, 344–349. https://doi.org/10.1016/j.indcrop.2009.12.004

Raja, T., Anand, P., Karthik, M., & Sundaraj, M. (2017). Evaluation of mechanical properties of natural fiber reinforced composites—A review. *International Journal of Mechanical Engineering and Technology*, *8*(7), 915–924.

Rana, A. K., Potluri, P., & Thakur, V. K. (2021). Cellulosic grewia optiva fibers: Towards chemistry, surface engineering and sustainable materials. *Journal of Environmental Chemical Engineering*, *9*(5), 106059. https://doi.org/10.1016/j.jece.2021.106059

Sakata, I., Morita, M., Furuichi, H., & Kawaguchi, Y. (1991). Improvement of plybond strength of paperboard by corona treatment. *Journal of Applied Polymer Science*, *42*(7), 2099–2104. https://doi.org/10.1002/app.1991.070420738

Shah, A. N., & Lakkad, S. C. (1981). Mechanical properties of jute-reinforced plastics. *Fiber Science and Technology*, *15*(1), 41–46. https://doi.org/10.1016/0015-0568(81)90030-0

Shah, D. U., Porter, D., & Vollrath, F. (2014). Can silk become an effective reinforcing fiber? A property comparison with flax and glass reinforced composites. *Composites Science and Technology*, 173–183.

Sinha, E., & Panigrahi, S. (2009). Effect of plasma treatment on structure, wettability of jute fiber and flexural strength of its composite. *Journal of Composite Materials*, *43*, 1791–1802.

Sreekala, M. S., Kumaran, M. G., Joseph, S., Jacob, M., & Thomas, S. (2000). Oil palm fiber reinforced phenol formaldehyde composites: Influence of fiber surface modifications on the mechanical performance. *Applied Composite Materials*, *7*(5–6), 295–329. https://doi.org/10.1023/A.1026534006291

Summerscales, J. (2021). A review of bast fibers and their composites : Part 4 ~ organisms and enzyme processes. *Composites Part A*, *140*, 106149. https://doi.org/10.1016/j.compositesa.2020.106149

Taj, S., & Munawar, M. A. (2014, October). *Natural fiber-reinforced polymer composites*. https://www.researchgate.net/profile/Munawar-Munawar-2/publication/228636811_Natural_fiber-reinforced_polymer_composites/links/544e8ced0cf29473161be3d9/Natural-fiber-reinforced-polymer-composites.pdf

Wang, B., Panigrahi, S., Tabil, L., & Crerar, W. (2007). Pre-treatment of flax fibers for use in rotationally molded biocomposites. *Journal of Reinforced Plastics and Composites*, *26*(5), 447–463. https://doi.org/10.1177/0731684406072526

Wu, X., & Dzenis, Y. A. (2006). Droplet on a fiber: Geometrical shape and contact angle. *Acta Mechanica*. https://doi.org/10.1007/s00707-006-0349-0

Xie, Y., Hill, C. A. S., Xiao, Z., Militz, H., & Mai, C. (2010). Silane coupling agents used for natural fiber/polymer composites: A review. *Composites Part A: Applied Science and Manufacturing*, *41*(7), 806–819. https://doi.org/10.1016/j.compositesa.2010.03.005

Yan, L., Chouw, N., & Yuan, X. (2012). Improving the mechanical properties of natural fiber fabric reinforced epoxy composites by alkali treatment. *Journal of Reinforced Plastics and Composites*, *31*(6), 425–437. https://doi.org/10.1177/0731684412439494

Yao, F., Wu, Q., Lei, Y., Guo, W., & Xu, Y. (2008). Thermal decomposition kinetics of natural fibers: Activation energy with dynamic thermogravimetric analysis. *Polymer Degradation and Stability*, *93*(1), 90–98. https://doi.org/10.1016/j.polymdegradstab.2007.10.012

Vinay, M., & Anshuman, S. (2014). Epoxy/wood apple shell particulate composite with improved mechanical properties. *International Journal of Engineering Research and Applications*, *4*(8), 142–145.

Zahran, M. K., Rehan, M. F., & El-Rafie, M. H. (2005). Single bath full bleaching of flax fibers using an activated sodium chlorite/hexamethylene tetramine system. *Journal of Natural Fibers*, *2*(2), 49–67. https://doi.org/10.1300/J395v02n02_04

Zin, M. H., Abdan, K., Mazlan, N., Zainudin, E. S., & Liew, K. E. (2018). The effects of alkali treatment on the mechanical and chemical properties of pineapple leaf fibers (PALF) and adhesion to epoxy resin. *IOP Conference Series: Materials Science and Engineering*, *368*(1). https://doi.org/10.1088/1757-899X/368/1/012035

2 Fabrication and Characterization of Green Composite

Shashikant Verma, Brijesh Gangil,
Ashutosh Gupta, Jitendra Verma, and Vishal Arya

CONTENTS

2.1 INTRODUCTION

Humans have been using materials/minerals from ancient times. Initially, they used natural materials: hunting, they used stone, sharp-edged wooden items; for living they

DOI: 10.1201/9781003272625-2

used mud houses. As the usefulness of the material increased, the material was used to make wood wheels and make iron arrows for easier life. Materials have played a vital part in human life from ancient to modern times and have contributed significantly to the modern world's economic progress. From pure metal to metal matrix, composites have made it sound in strength, light in weight, and easy to machine. In the case of wood, composites have made plywood to increase strength and make it easy to create the complicated shape. In the case of ceramics, high strength granite stone and tiles, etc. are examples of how the materials make life easy and modern (John & Thomas, 2008). Three broad groups of materials have been established: metals, ceramics, and polymers (Ahmad et al., 2015; Buschow, 2001). Category of composite is also based on metal, polymer, and ceramics matrix. A polymer from petroleum is known as synthetic, and another from a natural source is known as biopolymer, and its composites are emerging materials for the present time (Herrera-Franco & Valadez-González, 2004). Polymer composite has two primary constituents; one is polymer matrix materials, and another is reinforcement materials. A composite material gives better properties as compared to its parent's materials (Shibata et al., 2003). The material in the continuous phase inside the composite is called matrix material, and another material in the discontinuous and disperses phase is called reinforcement materials (Mukhopadhyay & Fangueiro, 2009). Polymer matrix materials are two types: one is thermosetting, and the other one is thermoplastic material.

Thermosetting polymers are those polymers that are not used again when the life of the material ends, but thermoplastic polymers are reusable materials (Oksman et al., 2003). Epoxy, Vinaylester, polyester, polyurethane, Bakelite, etc., are examples of thermosetting polymer and acrylics; A.B.S., nylons, P.V.C., and polyethylene are examples of thermoplastics (Bismarck et al., 2005). Reinforcing materials that provide the strength of polymer composite are fiber-filled or particulate-filled. If the polymer composites are fiber-reinforced, they also have synthetic fiber and natural fiber-filled. Synthetic fibers are glass fiber, Cardan fiber, and aramid fiber, while jute fiber, kenaf, sisal, oil palm, banana, and bamboo fibers are cellulose/lignocellulose natural fibers (Gangil et al., 2020). Polymer composite includes Al_2O_3 and other particulate filled (Hu et al., 2016); random orientated glass fiber reinforced (Kaundal et al., 2018); another synthetic fiber-reinforced composite (Tejyan et al., 2020); continuous carbon fiber and other continuous synthetic fiber-reinforced (Gangil et al., 2013); jute fiber and other natural fiber-reinforced thermosetting and thermoplastic polymer composites are the example of polymer composite. For the application of polymer composite, it is necessary to have a material having sufficient properties. Properties of polymer composite depend upon matrix material/fiber properties and the amount of matrix material/synthetic fiber. Thus, the properties of composite depend upon the constituent's materials, amount, chemical nature, orientation, and interfacial adhesion between the matrix and reinforcements materials (Arifur Rahman et al., 2015).

There are several applications of polymer composites such as in automobiles, aerospace electronics, marine industry, appliances, and consumer products; however, the problem of the use of the polymer composite is liquidating, and after the product has reached the end of its life, it is recycled and reused (Potluri & Chaitanya Krishna, 2019). Furthermore, the thermosetting polymer/synthetic fiber-reinforced composite necessitates a massive consumption of non-renewable oil-based resources. As a result, it is continuously reducing our natural resources and increasing the cost of production of fossil fuels. However, thermoplastic/natural fiber reinforced polymer composites are equal in properties and can recycle after the end of life. Scientists and researchers are consistently exploring alternate renewable, sustainable, and environment-friendly options (Potluri, 2019; Biagiotti et al., 2004).

In the view of this concept of "green composite" is an alternate option which gaining more and more attention from scientists and researchers, now a day because of environmental problem, reduction of the natural reserve. The matrix material in this green composite is reusable, and the reinforcement material is a biodegradable natural fiber. To control the environmental issues, green composite is rendered to resolve from the non-degradable polymer and synthetic fiber in the 21st century (Sathishkumar et al., 2013). Green composite is emerging to produce eco-friendly material, which has a sound impact on commercial and engineering applications. Present-day a due to the reduction of petroleum-based resources, there is a considerable requirement for renewable material which fulfill by green composite. Nevertheless, green composites have a very high requirement from global markets (Thakur et al., 2013; Väisänen et al., 2017).

2.2 APPLICATIONS OF GREEN COMPOSITE

As environment-friendly, biodegradable, economical, and simple in fabrication, the green composite having very waste are of application. Due to high specific properties, green composites are preferred as engineering materials. The utilization of green composites is in different industries, as discussed in Table 2.1.

TABLE 2.1
Applications of Natural Fiber in Different Industries

Application area	Material used	Applications
Construction	Basalt and Hemp Fiber	High Strength Panels
	Bamboo and Glass Fiber	High Strength Panels
	Short Sisal and Jute Fibers	False Celling's in Offices and Homes
	Jute Based Composite	Low-Cost House
	Coir and Polyester	Post Boxes, Roofing Panels
	Wood and Glass Fiber	Construction Structures
Electronics Industry	Kenaf Fiber with Polyester	N701ieco Mobile
	Bamboo Fiber	Outer Casing and Chassis of Dell
	Hemp and Glass	Server Hosting Racks in Data Centers
		Projector Casing and Other
	Banana Fiber/Polyester and Epoxy	Electric Equipment
Sporting Industry	Flax and Carbon Fibers	Race Bikes/Bicycle Frame
	Hemp and Flax Fiber	Racing Bicycle Spares
	Hemp and Glass Fibers	Skating Boards and Snowboards

(Continued)

TABLE 2.1 *(Continued)*
Applications of Natural Fiber in Different Industries

Application area	Material used	Applications
Transport Industry	Bamboo and Glass	Bodies of Trucks, Buses, Cars, and Rail
	Glass and Wood	Bodies of Trucks, Buses, Cars, and Rail
	Coir Fiber	Seats for the Rails, Trucks, Buses
	Flax Fibers Reinforced	Catamaran Boat Hulls
Energy Industry	Flax Fiber Reinforced	Blades of Turbine
	Bamboo Fiber	Blades of Turbine
		Shafts and Other Mechanical
	Hemp-Based Hybrid and Pure	Components
	Composites, Jute Composites	
Automobile Industry	Sisal, Kenaf, Hemp, Coir, Jute and	Automotive Components
	Sisal	Interior of Cars
	Coir Fiber	Side Covers, Mirror Covering, and Light
	Sisal and Roselle	Covers
	Kenaf and Glass Fibers	Bumpers of the Modern Cars
	Banana and Glass Fibers	Bumpers of the Modern Cars
	Hemp Fiber	Ford Motors Body Parts
	Cotton Fiber	Trabant
Aerospace Industry	Balsa Wood-Based Sandwich	Haviland Mosquitos and Albatros
	Flax Fiber-Based	Haviland Mosquitos and Albatros
	Hemp Fiber Reinforced	Electronics Placement Devices
	Basalt	Structural Components for the Aerospace

Source: Väisänen et al. (2017); Verma et al. (2019)

2.3   ADVANTAGES OF GREEN COMPOSITES

2.3.1   Advantages of Natural Fiber Over Synthetic Fiber

Natural fiber replaces the synthetic fiber in numerous ways due to low density, acceptable weight ratio, low energy requirement, and easy availability, and most essential is its bio-degradable nature. In comparison with cost, glass and carbon fiber range 1200 to 1800 US$/Ton and natural fiber varies 200 to 1000 US$/Ton (Ahmad et al., 2015; Faruk et al., 2014). A few highlights elaborate on the advantage of natural fiber over synthetic fiber.

- Low-density natural fiber makes 30% lighter products.
- Easier recycling and processing of natural fiber over synthetic fiber
- Mechanical properties of natural fibers are quite acceptable as compared to synthetic fiber.
- Good surface finish makes the product attractive.
- Reduced problem of disposing of the product after its use has ended

2.3.2   ADVANTAGE OF THERMOPLASTIC OVER THERMOSETTING

First of all, the difference between thermoplastic and thermosetting is that thermo-plastics can be heated and molded again and again whereas thermosetting once solid-ified cannot be remelted and reshaped. When biodegradable resin, petrochemical plastics release carbon when they are disposed of, do not hold/release carbon during casting and disposal. Biodegradable plastics are disposed of quickly compared to thermosetting plastics, so it is vital for a healthy environment (Azwa et al., 2013; Shanks et al., 2004).

2.3.3   THE DISADVANTAGE OF GREEN COMPOSITE

Apart from the amazing advantage of green composite there are a few drawbacks from the synthetic polymer composite. There are a few disadvantages as discussed next:

- Due to low density and weight, it produced unstable bonding.
- It produced lower impact strengths properties.
- Good moisture absorption causes swelling of fibers which reduces the durability.
- The source of green composite is natural, so it requires some chemical treat-ment before fabrication.

(Sanjay et al., 2015)

2.4   CONSTITUTES OF GREEN COMPOSITE

Green composites are formed by natural fiber and biodegradable resin, in which the natural fiber reinforces biodegradable resin. Minerals, animals, and plants fibers are the three primary groups of natural fibers. Plan fibers are vegetable fiber with cellu-lose, hemicelluloses, and lignin. The source of plant fibers is grass, seed, leaf, wood, and bast. **Figure 2.1** shows a comprehensive classification of natural fiber (Verma et al., 2019).

2.4.1   BIODEGRADABLE RESIN

Most of the time, the recycling product industries uses thermoplastic material. In which polyethylene (P.E.), polyvinyl chloride (P.V.C.), and polystyrene (P.S.) is used for packaging industries. A few thermoplastics are listed next (Frone et al., 2011):

- Polycarbonate
- Acetal Copolymer Polyoxymethylene
- Acetal Homopolymer Polyoxymethylene
- Acrylic
- Nylon
- Polyethylene
- Polypropylene
- Polystyrene
- Polyvinyl chloride (PVC)
- Teflon

Natural fiber

Animal
Silk
Wool
Hair

Mineral
Asbestos

Plants fiber

Bast	Leaf	Seed	Fruits	Wood	Stalk	Grass
Jute	Sisal	Loofah	Oil Palm	Hard	Rice	Bamboo
Remie	Banana	Milkweed	Coil	Wood	Wheat	Bagasse
Flax	Henequen	Kapok		Soft wood	Barley	Corn
Kenaf	Agave	Cotton			Maize	Sabai
Roselle	Palf				Oat	Rape
Mesta	Abaca				rye	Esparto
Hemp						Cancry

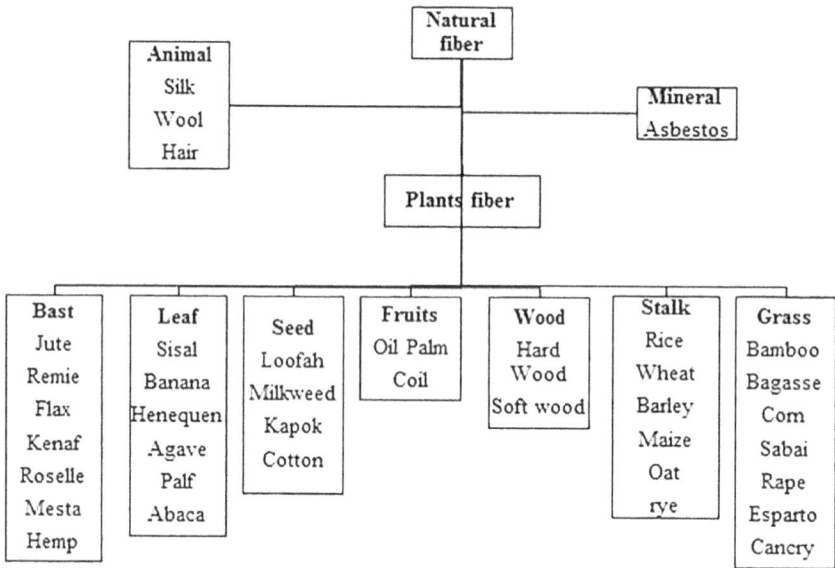

FIGURE 2.1   Classification of green composite.

Source: Verma et al. (2019)

Beyond the thermoplastic material, if we talk about biodegradable material, there are five biodegradable materials as listed next:

- Thermoplastic Starch-Based Plastics (T.P.S.)
- Polyhydroxyalkanoates (P.H.A.)
- Polylactic Acid (P.L.A.)
- Polybutylene Succinate (PBS)
- Polycaprolactone (P.C.L.)

Among these biodegradable materials, T.P.S. are inexpensive and readily available materials used in food packaging/take-out foods and one-time-use plastics. P.L.A. is made from sugar cane and cornstarch and is used to manufacture large consumer products and medical strength implementation rods and screws (Madhavan Nampoothiri et al., 2010).

2.5   FABRICATION OF GREEN COMPOSITE

The process of converting two or more renewable materials into final things is referred to as a green composite fabrication. Natural fiber with one biopolymer and maybe two biopolymers, or recycled polymer, may be used in the finished product (Ben & Kihara, 2007; George et al., 2010). Biopolymers are mainly formed from sugar, starch, protein, and vegetable oil, while natural fibers are carbohydrates or plant fiber (Ben & Kihara, 2007). There are many fabrication processes for making green composites, but the

MATERIAL SELECTION	⇨	MATERIAL TREATMENT	⇨	DRYING	⇨	SEMI FINISHED PRODUCT	⇨	DRYING	⇨	PRODUCT

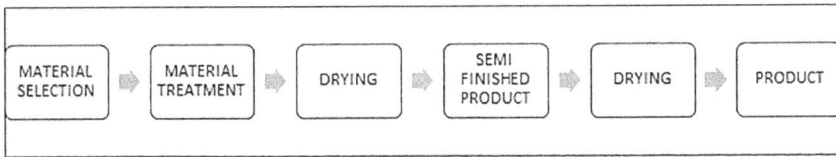

FIGURE 2.2 Flow diagram of green composites.

fabrication of green composite is completed in four steps. The steps necessary for the fabrication of green composites are given next and also shown in **Figure 2.2**:

- Material selection
- Material preprocessing
- Semifinished product manufacturing
- Green composite fabrication

Because the production of green composites differs from standard polymer composites, parameters such as morphological changes, surface contact, surface appearance, and chemical coupling between constituents must be studied before manufacture. The morphology, thermal instability, and water absorptions inside the natural fiber all affect green composites. Biopolymers, on the other hand, such as P.L.A., are crystalline and have poor flow characteristics, as well as rapid deterioration above 190°C (Lim et al., 2008). Natural fibers differ in their hydrophilic characteristics, which limit fiber-matrix adhesion (Thakur et al., 2013). Natural fibers from different sources having different chemical compositions showing different morphology are also a big challenge for fabricating green composites (Faruk et al., 2012).

2.5.1 MATERIAL SELECTION

Before material fabrication, there is the necessary material selection like the cost of fiber, functional properties of fiber, and processability of green composite. Low processing costs affect overall product costs (Tazi et al., 2016). Mechanical, physical, thermal, and tribological properties depend upon the functional properties of raw material used in green composites (George et al., 2010). Viscosity, specific heat, thermal conductivity, and crystallinity also affect the processability of green composites, which are also considered at the time of material selections (Menzel et al., 2013).

2.5.2 PREPROCESSING

The key steps in the processing of green composites include fiber preparation, drying, and fiber treatments. Fibers can be classified as plant fiber, animal fiber, or mineral fiber, and then separated into cellulose, hemicellulose, and lignin, depending on the source of natural fiber (Sain & Pervaiz, 2008). Among these fibers, cellulose fibers are the most commonly employed in the manufacture of polymer composites. The most important process for green composites after fiber preparation is drying. During the drying process, humidity is an important factor to consider. In their study, (Ayrilmis & Jarusombuti, 2011; La Mantia & Morreale, 2011) recommended that drying natural fiber to 1–3% humidity and drying biopolymer to less than 1% humidity was required prior to processing. Drying time and temperature are also notable and studied for each fiber and a biopolymer. Fiber treatment is also necessary before final casting due to impurities present in the fiber, which reduces chemical bonding between fiber and biopolymer. Various pretreatment processes like physical and chemical processes are used for fiber treatment: in physical treatment, homogenization and vacuum treatment; and in chemical treatments, fiber dip in the alkali, acid hydrolysis, silanes, and isocyanates solution for a particular time. As discussed earlier, after chemical treatment, fiber is passed out by the drying process.

2.5.3 SEMIFINISHED PRODUCT MANUFACTURING

Semifinished products are prepared before going to the final product. For compression molding, fibers and biopolymers are converted in granules, sheets, prepregs, or laminates form, but for thermoforming process, raw material is converted in sheets form. Fiber mat technology and compounding technology are used for long fiber (10–30mm) and short fiber (< 10mm) respectively to convert into the semifinished product (Faruk et al., 2012; Reddy & Yang, 2011).

2.6 GREEN COMPOSITE FABRICATION PROCESS

2.6.1 COMPRESSION MOLDING

The primary components in the compression molding process are pressure and temperature, in which the material is heated in the mold cavity under heat and pressure with solidification time/cooling time followed by part removal from the mold, as indicated in **Figure 2.3**.

Few studies have used prepreg sheets as an alternate stacking method, in which alternate arrangements of natural fiber mats followed by thermoplastic resign are used and applied heat and pressure (Du et al., 2010). In their study, (Ben & Kihara, 2007) used an alternate stacking method to fabricate kenaf fiber with P.L.A. sheets. The melting process was done for 10 MPa pressure for 10 seconds and holding 1MPa for 20 minutes. The temperature is 185° maintained throughout the process.

In their investigations for the manufacture of cellulose and P.L.A. sheets, (Singh et al., 2020) used two steps: pre-pressing and pressing. The platen temperature was kept at 165°C during pre-pressing, with 5 MPa pressure applied for three minutes and 15 MPa pressure applied for the rest of the time. In their studies, (Reddy & Yang, 2011) used jute and wheat gluten to make composite sheets, in which pressure was applied for 140 MPa for five to 20 minutes time, and 150–180ºC temperature was then cooled by cold water followed by removal of sheets. Jute fiber and epoxy repair sheets formed by hand lay-up technique followed by compression molding process curing temperature at 80–130°C were studied by (Zampaloni et al., 2007), which looked at the effect of curing temperature on mechanical properties of composites.

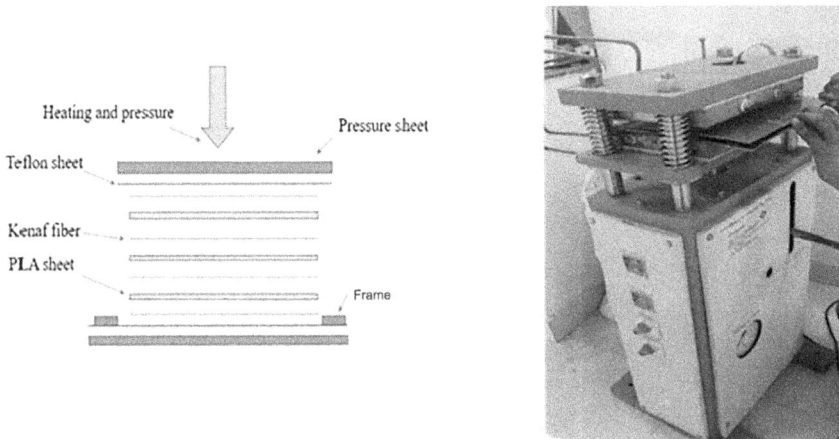

FIGURE 2.3   Compression molding for kenaf and PLA by alternat stacking process.

Source: Ben & Kihara (2007)

Curing time, temperature, and pressure are critical parameters in the compression molding process that must be tuned to achieve the desired and best results. (Benthien & Thoemen, 2012) evaluated the influence of press temperature on panel qualities using wood flour and polypropylene, and the optimal press temperature was 210°C.

2.6.2 THERMOFORMING

The thermoforming process required a dry semifinished sheet to avoid edged defects after heating. A clamping device is required to clamp the sheet to avoid twisting. Convection heaters on one sheet side heat the sheets to softening temperature. On the other side, one cavity of the desired shape is fixed, and then vacuum pressure is applied for forming process, as shown in **Figure 2.4**. (Kc et al., 2015; Lim et al., 2008) prepared a sheet by the thermoforming process of 17% laminates of cellulose and P.L.A. sheets and suggested the thermoforming temperature rank 80–110°C. Another study by (Zampaloni et al., 2007) prepared kenaf and polypropylene sheets by thermoforming process and optimized the thermoforming temperature is 190°C, heating time 15 minutes, and drown depth is 50.8mm.

2.6.3 EXTRUSION

P.L.A. matrix and natural fiber composite widely fabricated by extrusion process (shown in **Figure 2.5**, an industrial unit). The advantage of this process is compounding ability and functional versatility. The raw materials are prepared as blinding proportions per machine requirement (single-screw and twin-screw extrusion). There is variation in temperature as 185, 175, 155, and 140°C with variable speeds of up to 150 rpm in the extrusion process. At the end of the extrusion process samples were

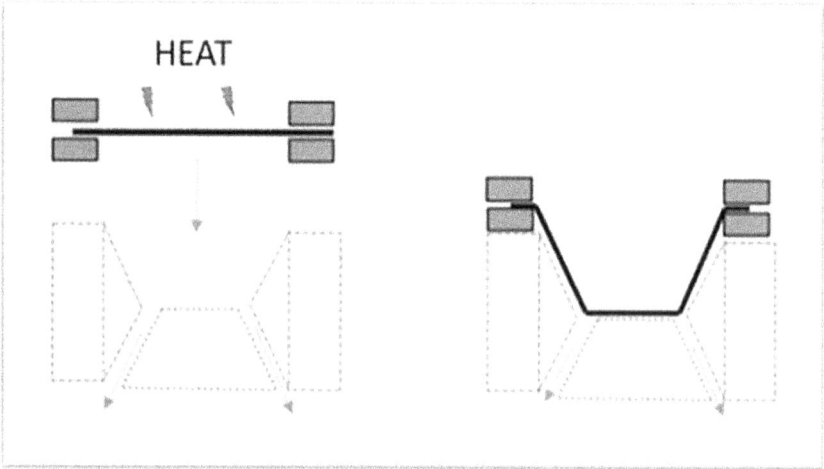

FIGURE 2.4   Systematic diagram of thermoforming process.

Source: Kc et al. (2015)

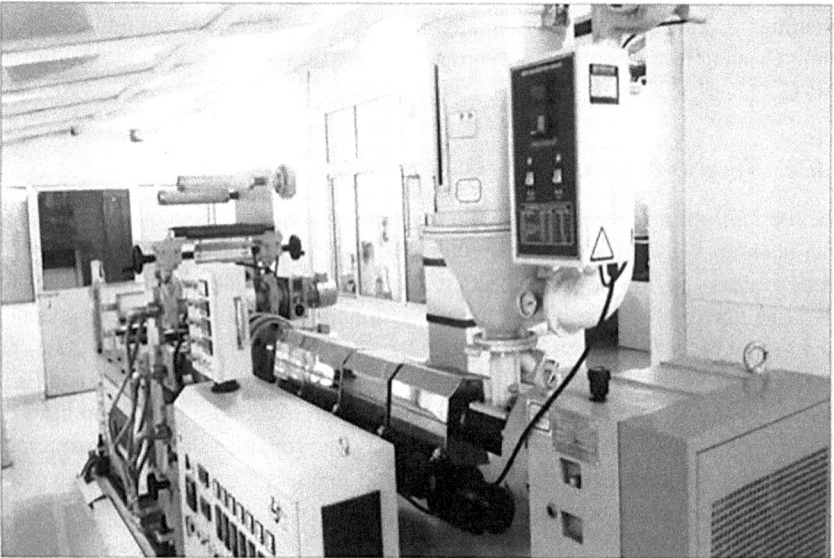

FIGURE 2.5   Industrial extrusion machine.

cut with the pelletizer and finally finished parts were dried. In (Atiqah et al., 2014), the first fiber was treated in NaOH solution at 50°C for 24 hours, then washed in water and dried again at 60°C. The various components are blended at high speed and make a homogeneous solution. The last process to extrusion was directed at the speed of the rotor.

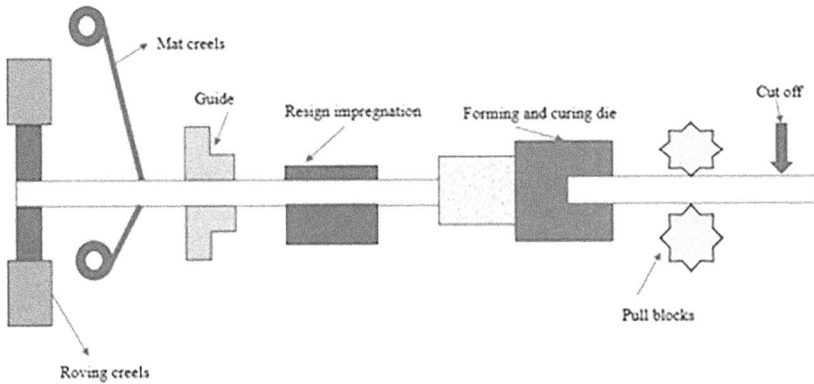

FIGURE 2.6 Systematic block diagram of pultrusion process.

Source: Fairuz et al. (2014)

2.6.4 PULTRUSION

Pultrusion is similar to extrusion; however, the materials are dragged rather than pushed in the pultrusion process shown in **Figure 2.6**. In the pultrusion process, the treated continuous fiber is taped in creels attached by roving creels, as shown in the figure. The continuous fiber was pulled and passed through the resin bath, and the final material shape passed through the die cross-section. The final product depends upon the die, either in the form of rods or bars. (Van De Velde & Kiekens, 2001) developed a material by the pultrusion process. The major advantage of this process is to produce continuous materials with a constant cross-section profile of material (Fairuz et al., 2014).

2.6.5 RESIN TRANSFER MOLDING

The resin transfer molding process is used to fabricate continuous shape material with woven mat/discontinuous fiber mat as shown in **Figure 2.7**. This process's benefit is to control fiber orientation and fiber orientation useful for resin transfer. (Liu et al., 2014) made unidirectional abaca fiber that was fabricated using resin transfer molding process. (Oksman et al., 2002) in their study used sisal as unidirectional fiber and epoxy as resin and examined the material's mechanical properties. (Oksman, 2001) makes high-quality flax composite by resin transfer molding process.

2.6.6 VACUUM-ASSISTED RESIN TRANSFER MOLDING (VARTM)

Vacuum-assisted resin transfer molding process is based on negative pressure generated by the vacuum pump. There are two sides, one high-low pressure and another that is low-pressure; due to pressure difference the resin flows from low pressure to high pressure. Firstly surface is cleaned by acetone solution to remove dust from it and apply silica gel on the mold surface to avoid the sticking behavior of composites

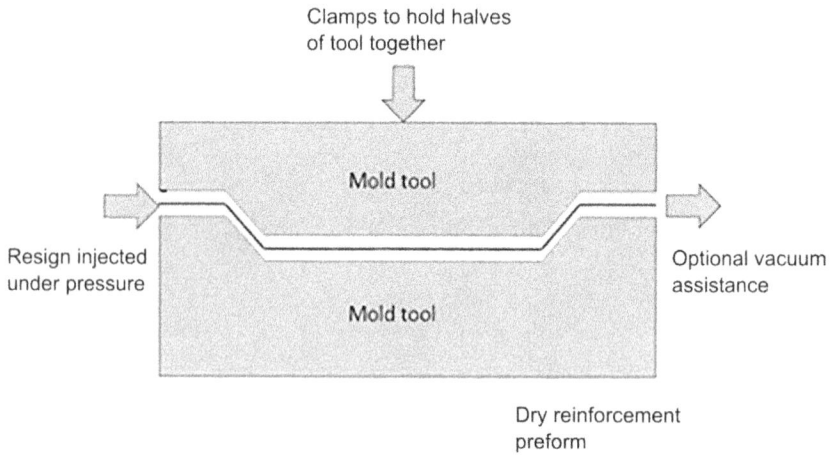

FIGURE 2.7   Systematic diagram of resin transfer molding process.

Source: Liu et al. (2014)

to the mold surface (Chaudhary et al., 2019). Ten layers of natural fiber are staged together on the plain surface covered by peel ply (work as separation agent from composite to resin infusion mesh), followed by resin infusion mesh (use for uniform flow of resin) finally by vacuum bagging. **Figure 2.8** shows systemic arrangements of vacuum-assisted resin transfer molding process.

Resin inlet and outlet pipe were connected by vacuum bag with the help of sealant tape. The vacuum pump was turned on; the entire air was empty from the mold, then the resin flow stared in the mold uniformly until the point resin flow to the entire mold was achieved. Mold was left for solidification for 24 hours at room temperature (Verma et al., 2022). (Maslinda et al., 2017) used cellulose fibers and cast composite by VARTM process. (Choudhary et al., 2019) fabricated composite by VARTM process and compared with hand lay-up process.

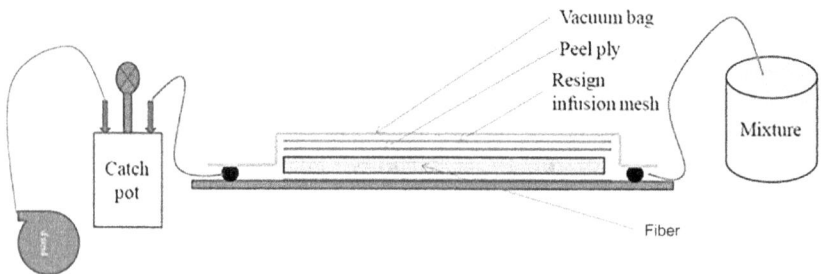

FIGURE 2.8   Systematic diagram of vacuum-assisted resin transfer molding process.

Source: Verma et al. (2022)

Mold releasing spry

Resign   Impregnation roller

25 kg weight application for curing

weight

Composite ready after 24 hours

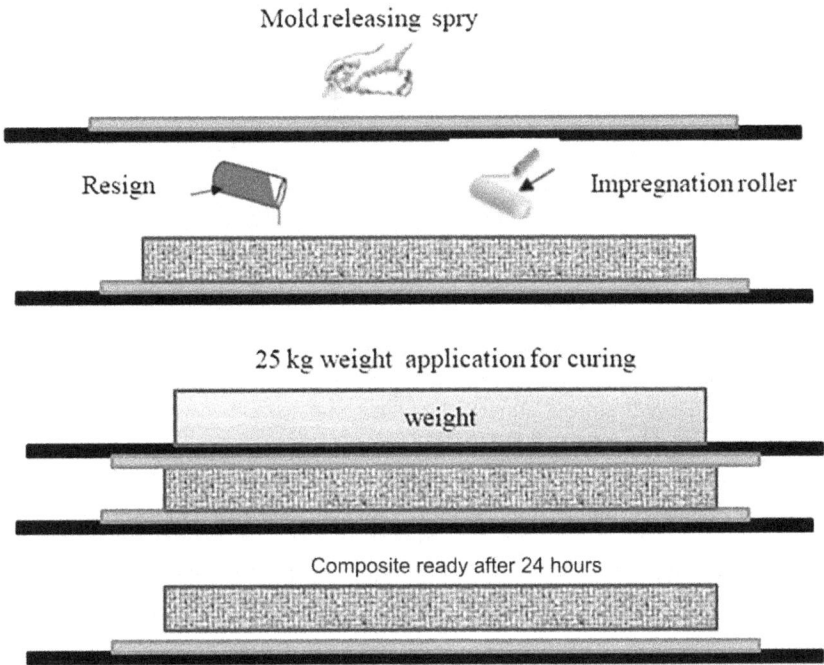

FIGURE 2.9   Systematic hand lay-up process.

Source: Verma et al. (2022)

2.6.7   HAND LAY-UP PROCESS

The most effortless process for casting composite is the hand lay-up process. First, the mold was cleaned with acetone to remove dust particles on the surface and then applied the mold releasing agent. After that, resin, and the fiber layers are applied one by one. The steps of the hand lay-up process are shown in **Figure 2.9**. At last, mold was expelled to solidification for 24 hours at atmospheric air temperature. After solidification, the samples are removed from the mold and ready for testing. Many researchers have worked by using the hand lay-up process (Singh et al., 2019; Verma et al., 2019).

2.7   CHARACTERIZATION OF GREEN COMPOSITE

Many factors that affect the properties of natural fibers depend upon the fiber's origin, location of natural fiber, quality, weather, orientation, and effect by the chemical structure, cellulose contents. Compared with synthetic fiber, natural fiber has fewer tensile properties. In this section, different properties are discussed that affect the properties of composite applications of synthetic fiber composite instead of green composite.

2.7.1 Physical and Mechanical Properties

It is vital to research the physical qualities of natural fiber, such as fiber diameters, flaws, strength, variability, crystallinity, and structure, before employing it. As compared to synthetic fiber, natural fiber possesses a lower density. Jute fiber, hemp, flux alfa, ramie, and cotton have a density of 1.4–1.5 g/cm³ which is less than glass fiber of 2.5g/cm³ (Sanjay et al., 2015). The density of glass fiber affects the mechanical properties of the composite compared with glass fiber, which is commonly used in polymer composite having lower tensile strength than green composite. Specific tensile strength (tensile strength/density) of some natural fiber is less than synthetic fiber (Gangil et al., 2020). Table 2.2 shows the difference between natural fiber's physical and mechanical properties compared with synthetic fiber.

The main contents in the natural fibers are cellulose, hemicellulose, and lignin which provide a continuous/discontinuous chain that provides strength and stiffness of the green composite (Gangil et al., 2013; Verma et al., 2019). Tensile strength and young's modulus of green composite depend upon the content of cellulose in the natural fiber. Cellulose increases the tensile strength, but hemicellulose decreases the green composite's tensile strength and young's modulus. The strength of the fiber is also affected by cellulose fibrils like parallel cross-linking. The stiffness of the composite also depends upon the cellulose content present in the fiber. Lignin content also affects the stiffness properties of natural fiber. The microfibrillar angle is also a major factor that affects the composite's mechanical properties. Stiffness of

TABLE 2.2
Physical and Mechanical Properties of Natural and Synthetic Fiber

Fiber	Density (g/cm³)	Failure strain (%)	Tensile strength (MPa)	Stiffness/ young's modulus (GPa)	Specific tensile strength (MPa/gcm³)	Specific young's modulus (GPa/gcm³)
E-Glass	2.5	2.5	2000–3000	70	800–1400	29
Aramide	1.4	3.3–3.7	3000–3150	63–67	2140–2250	-
Carbon	1.4	1.4–1.8	4000	230–240	2860	-
Ramie	1.4	2.0–3.8	400–938	44–128	250–620	29–85
Jute	1.4–1.5	1.5–1.8	393–800	10–55	300–610	7.1–39
Flax	1.4	1.2–3.2	345–1820	27–80	230–1220	18–53
Hemp	1.4	1.6	550–1110	58–70	370–740	39–47
Sisal	1.3–1.5	2.0–2.5	550–850	9.4–25	360–600	6.7–20
Alfa	1.4	1.5–2.5	188–308	18–25	134–220	13–18
Cotton	1.5–1.6	3–10	250–800	5.5–13	190–530	3.7–8.4
Silk	1.3	15–60	100–1500	5–25	100–1500	4–20
Wool	1.3	13–35	50–315	2.3–5	38–240	1.8–3.8
Coir	1.2	15–30	131–220	4–6	110–180	3.3–5

Source: Gangil et al. (2020)

composite is directly proportional to microfibrillar angle: if microfibril angle is less than 5° the stiffness is around to 50–80 GPa, and if it is greater than or between 40° and 50°, stiffness is reduced up to 20 GPa.

2.7.2 TRIBOLOGICAL PROPERTIES OF GREEN COMPOSITE

Nowadays, environmental issues are critical in some metro cities. So that the tribological properties are significant to analyze, there are two critical tribological phenomena during the relative motion of solid surfaces: friction and wear. The tribological performance and wear process of polymer composites have been extensively studied by many researchers (Klaus Friedrich, 2008; Nirmal et al., 2015). Wear is an extrinsic property of the polymer. It depends on various parameters like sliding speed, applied load, surface roughness, and temperature. In green composite reinforcement like fillers, micro or nano reinforcement increases wear resistance. Wear mass loss depends on the load and sliding speed (T. Singh et al., 2018). At low load and low sliding speed, the degree of penetration is less, while on increasing the load, the wear mass loss increases because a large number of abrasive particles are in action. The degree of penetration gets improved so that more deep grooves are formed. A large amount of material is removed due to deeper groves. The wear volume loss increases proportionally as the sliding speed and abrasive particle size increase. Depending on the materials, the coefficient of friction decreases as the sliding speed increases (Kumar et al., 2019).

2.8 CONCLUSION

Most bulk material used in the automotive sector, space industries, and sports sector hurts humans and the environment. In terms of using green composite, many industrial sectors replace synthetic materials with natural materials. Greater environmental awareness can increase regulations on pollution, product end life disposal, productivity, and green innovation. The development of green industries includes the development of eco-friendly material, and its supply reduces adverse environmental effects and improves recycling of green material. Such efforts reduce wastage, and decrease manufacturing costs and hazardous pollution. Meanwhile, green composite is a step towards long-term techno-economic progression and ecological development.

REFERENCES

Ahmad, F., Choi, H. S., & Park, M. K. (2015). A review: Natural fiber composites selection in view of mechanical, light weight, and economic properties. *Macromolecular Materials and Engineering*, *300*(1), 10–24. https://doi.org/10.1002/mame.201400089

Alexander, B., & Supriya Mishra, T. L. (2005). Plant fibers as reinforcement for green composites. In Alexander, B., Supriya Mishra, T. L., & Thomas, L. (Eds.), *Natural Fibers, Biopolymers, and Biocomposites* (pp. 52–128). London: CRC Press. https://doi.org/10.1201/9780203508206-10

Arifur Rahman, M., Parvin, F., Hasan, M., & Hoque, M. E. (2015). Introduction to manufacturing of natural fibre-reinforced polymer composites. *Manufacturing of Natural Fibre Reinforced Polymer Composites*, 17–43. https://doi.org/10.1007/978-3-319-07944-8_2

Atiqah, A., Maleque, M. A., Jawaid, M., & Iqbal, M. (2014). Development of kenaf-glass reinforced unsaturated polyester hybrid composite for structural applications. *Composites Part B: Engineering*, *56*, 68–73. https://doi.org/10.1016/j.compositesb.2013.08.019

Ayrilmis, N., & Jarusombuti, S. (2011). Flat-pressed wood plastic composite as an alternative to conventional wood-based panels. *Journal of Composite Materials*, *45*(1), 103–112. https://doi.org/10.1177/0021998310371546

Azwa, Z. N., Yousif, B. F., Manalo, A. C., & Karunasena, W. (2013). A review on the degradability of polymeric composites based on natural fibres. *Materials and Design*, *47*, 424–442. https://doi.org/10.1016/j.matdes.2012.11.025

Ben, G., & Kihara, Y. (2007). Development and evaluation of mechanical properties for kenaf fibers/PLA composites. *Key Engineering Materials*, *334–335*, 489–492. https://doi.org/10.4028/WWW.SCIENTIFIC.NET/KEM.334-335.489

Benthien, J. T., & Thoemen, H. (2012). Effects of raw materials and process parameters on the physical and mechanical properties of flat pressed WPC panels. *Composites Part A: Applied Science and Manufacturing*, *43*(4), 570–576. https://doi.org/10.1016/j.compositesa.2011.12.028

Biagiotti, J., Puglia, D., & Kenny, J. M. (2004). A review on natural fibre-based composites—Part I: Structure, processing and properties of vegetable fibres. *Journal of Natural Fibers*, *1*(2), 37–68. https://doi.org/10.1300/J395v01n02_04

Buschow, K. (2001). *Encyclopedia of Materials: Science and Technology*. Oxford: Elsevier.

Choudhary, M., Sharma, A., Dwivedi, M., & Patnaik, A. (2019). A Comparative study of the physical, mechanical and thermo-mechanical behavior of GFRP composite based on fabrication techniques. *Fibers and Polymers*, *20*(4), 823–831. https://doi.org/10.1007/s12221-019-8863-6

Du, Y., Zhang, J., Yu, J., Lacy, T. E., Xue, Y., Toghiani, H., Horstemeyer, M. F., & Pittman, C. U. (2010). Kenaf bast fiber bundle-reinforced unsaturated polyester composites. IV: Effects of fiber loadings and aspect ratios on composite tensile properties. *Forest Products Journal*, *60*(7–8), 582–591. https://doi.org/10.13073/0015-7473-60.7.582

Fairuz, A. M., Sapuan, S. M., Zainudin, E. S., & Jaafar, C. N. A. (2014). Polymer composite manufacturing using a pultrusion process: A review. *American Journal of Applied Sciences*, *11*(10), 1798–1810. https://doi.org/10.3844/ajassp.2014.1798.1810

Faruk, O., Bledzki, A. K., Fink, H. P., & Sain, M. (2012). Biocomposites reinforced with natural fibers: 2000–2010. *Progress in Polymer Science*, *37*(11), 1552–1596. https://doi.org/10.1016/j.progpolymsci.2012.04.003

Faruk, O., Bledzki, A. K., Fink, H. P., & Sain, M. (2014). Progress report on natural fiber reinforced composites. *Macromolecular Materials and Engineering*, *299*(1), 9–26. https://doi.org/10.1002/mame.201300008

Frone, A. N., Berlioz, S., Chailan, J. F., Panaitescu, D. M., & Donescu, D. (2011). Cellulose fiber-reinforced polylactic acid. *Polymer Composites*, *32*(6), 976–985. https://doi.org/10.1002/PC.21116

Gangil, B., Patnaik, A., & Kumar, A. (2013). Mechanical and wear behavior of vinyl ester-carbon/cement by-pass dust particulate filled homogeneous and their functionally graded composites. *Science and Engineering of Composite Materials*, *20*(2), 105–116. https://doi.org/10.1515/secm-2012-0109

Gangil, B., Ranakoti, L., Verma, S., Singh, T., & Kumar, S. (2020). Natural and synthetic fibers for hybrid composites. *Hybrid Fiber Composites*, 1–15. https://doi.org/10.1002/9783527824571.ch1

George, G., Joseph, K., Boudenne, A., & Thomas, S. (2010). Recent advances in green composites. *Key Engineering Materials*, *425*, 107–166. https://doi.org/10.4028/www.scientific.net/KEM.425.107

Herrera-Franco, P. J., & Valadez-González, A. (2004). Mechanical properties of continuous natural fibre-reinforced polymer composites. *Composites Part A: Applied Science and Manufacturing, 35*(3), 339–345. https://doi.org/10.1016/j.compositesa.2003.09.012

Hu, Y., Du, G., & Chen, N. (2016). A novel approach for Al2O3/epoxy composites with high strength and thermal conductivity. *Composites Science and Technology, 124*, 36–43. https://doi.org/10.1016/j.compscitech.2016.01.010

John, M. J., & Thomas, S. (2008). Biofibres and biocomposites. *Carbohydrate Polymers, 71*(3), 343–364. https://doi.org/10.1016/j.carbpol.2007.05.040

Kaundal, R., Patnaik, A., & Satapathy, A. (2018). Mechanical characterizations and development of erosive wear model for Al2O3-filled short glass fiber-reinforced polymer composites. *Proceedings of the Institution of Mechanical Engineers, Part L: Journal of Materials: Design and Applications, 232*(11), 893–908. https://doi.org/10.1177/1464420716654307

Kc, B., Pervaiz, M., Faruk, O., Tjong, J., & Sain, M. (2015). Green composite manufacturing via compression molding and thermoforming. *Manufacturing of Natural Fibre Reinforced Polymer Composites*, 45–63. https://doi.org/10.1007/978-3-319-07944-8_3

Klaus Friedrich, A. K. S. (2008). Tribology of polymeric nanocomposites. *Tribology and Interface Engineering Series, 55*(C), ii. https://doi.org/10.1016/S1572-3364(13)70001-6

Kumar, S., Mer, K. K. S., Gangil, B., & Patel, V. K. (2019). Synergy of rice-husk filler on physico-mechanical and tribological properties of hybrid Bauhinia-vahlii/sisal fiber reinforced epoxy composites. *Journal of Materials Research and Technology, 8*(2), 2070–2082. https://doi.org/10.1016/j.jmrt.2018.12.021

La Mantia, F. P., & Morreale, M. (2011). Green composites: A brief review. *Composites Part A: Applied Science and Manufacturing, 42*(6), 579–588. https://doi.org/10.1016/J.COMPOSITESA.2011.01.017

Lim, L. T., Auras, R., & Rubino, M. (2008). Processing technologies for poly(lactic acid). *Progress in Polymer Science (Oxford), 33*(8), 820–852. https://doi.org/10.1016/j.progpolymsci.2008.05.004

Liu, K., Zhang, X., Takagi, H., Yang, Z., & Wang, D. (2014). Effect of chemical treatments on transverse thermal conductivity of unidirectional abaca fiber/epoxy composite. *Composites Part A: Applied Science and Manufacturing, 66*, 227–236. https://doi.org/10.1016/j.compositesa.2014.07.018

Madhavan Nampoothiri, K., Nair, N. R., & John, R. P. (2010). An overview of the recent developments in polylactide (PLA) research. *Bioresource Technology, 101*(22), 8493–8501. https://doi.org/10.1016/j.biortech.2010.05.092

Maslinda, A. B., Abdul Majid, M. S., Ridzuan, M. J. M., Afendi, M., & Gibson, A. G. (2017). Effect of water absorption on the mechanical properties of hybrid interwoven cellulosic-cellulosic fibre reinforced epoxy composites. *Composite Structures, 167*, 227–237. https://doi.org/10.1016/j.compstruct.2017.02.023

Menzel, C., Olsson, E., Plivelic, T. S., Andersson, R., Johansson, C., Kuktaite, R., Järnström, L., & Koch, K. (2013). Molecular structure of citric acid cross-linked starch films. *Carbohydrate Polymers, 96*(1), 270–276. https://doi.org/10.1016/j.carbpol.2013.03.044

Mukhopadhyay, S., & Fangueiro, R. (2009). Physical modification of natural fibers and thermoplastic films for composites—A review. *Journal of Thermoplastic Composite Materials, 22*(2), 135–162. https://doi.org/10.1177/0892705708091860

Nirmal, U., Hashim, J., & Megat Ahmad, M. M. H. (2015). A review on tribological performance of natural fibre polymeric composites. *Tribology International, 83*, 77–104. https://doi.org/10.1016/j.triboint.2014.11.003

Oksman, K. (2001). High Quality Flax Fibre Composites Manufactured by the Resin Transfer Moulding Process. *Journal of Reinforced Plastics and Composites, 20*(7), 621–627. https://doi.org/10.1177/073168401772678634

Oksman, K., Skrifvars, M., & Selin, J. F. (2003). Natural fibres as reinforcement in polylactic acid (PLA) composites. *Composites Science and Technology, 63*(9), 1317–1324. https://doi.org/10.1016/S0266-3538(03)00103-9

Oksman, K., Wallström, L., Berglund, L. A., & Toledo Filho, R. D. (2002). Morphology and mechanical properties of unidirectional sisal-epoxy composites. *Journal of Applied Polymer Science, 84*(13), 2358–2365. https://doi.org/10.1002/app.10475

Potluri, R. (2019). Natural fiber-based hybrid bio-composites: Processing, characterization, and applications. *Green Composites*, 1–46. https://doi.org/10.1007/978-981-13-1972-3_1

Potluri, R., & Chaitanya Krishna, N. (2019). Potential and applications of green composites in industrial space. *Materials Today: Proceedings, 22*, 2041–2048. https://doi.org/10.1016/j.matpr.2020.03.218

Reddy, N., & Yang, Y. (2011). Biocomposites developed using water-plasticized wheat gluten as matrix and jute fibers as reinforcement. *Polymer International, 60*(4), 711–716. https://doi.org/10.1002/pi.3014

Sain, M., & Pervaiz, M. (2008). Mechanical properties of wood—Polymer composites. *Wood-Polymer Composites*, 101–117. https://doi.org/10.1533/9781845694579.101

Sanjay, M. R., Arpitha, G. R., & Yogesha, B. (2015). Study on mechanical properties of natural—Glass fibre reinforced polymer hybrid composites: A review. *Materials Today: Proceedings, 2*(4–5), 2959–2967. https://doi.org/10.1016/j.matpr.2015.07.264

Sathishkumar, T. P., Navaneethakrishnan, P., Shankar, S., Rajasekar, R., & Rajini, N. (2013). Characterization of natural fiber and composites—A review. *Journal of Reinforced Plastics and Composites, 32*(19), 1457–1476. https://doi.org/10.1177/0731684413495322

Shanks, R. A., Hodzic, A., & Wong, S. (2004). Thermoplastic biopolyester natural fiber composites. *Journal of Applied Polymer Science, 91*(4), 2114–2121. https://doi.org/10.1002/app.13289

Shibata, M., Ozawa, K., Teramoto, N., Yosomiya, R., & Takeishi, H. (2003). Biocomposites made from short abaca fiber and biodegradable polyesters. *Macromolecular Materials and Engineering, 288*(1), 35–43. https://doi.org/10.1002/mame.200290031

Singh, J. I. P., Singh, S., & Dhawan, V. (2020). Effect of alkali treatment on mechanical properties of jute fiber-reinforced partially biodegradable green composites using epoxy resin matrix. *Polymers and Polymer Composites, 28*(6), 388–397. https://doi.org/10.1177/0967391119880046

Singh, T., Gangil, B., Patnaik, A., Kumar, S., Rishiraj, A., & Fekete, G. (2018). Physico-mechanical, thermal and dynamic mechanical behaviour of natural-synthetic fiber reinforced vinylester based homogenous and functionally graded composites. *Materials Research Express, 6*(2), 025704. https://doi.org/10.1088/2053-1591/AAEE30

Singh, T., Gangil, B., Singh, B., Verma, S. K., Biswas, D., & Fekete, G. (2019). Natural-synthetic fiber reinforced homogeneous and functionally graded vinylester composites: Effect of bagasse-Kevlar hybridization on wear behavior. *Journal of Materials Research and Technology, 8*(6), 5961–5971. https://doi.org/10.1016/j.jmrt.2019.09.071

Tazi, M., Sukiman, M. S., Erchiqui, F., Imad, A., & Kanit, T. (2016). Effect of wood fillers on the viscoelastic and thermophysical properties of HDPE-wood composite. *International Journal of Polymer Science, 2016*. https://doi.org/10.1155/2016/9032525

Tejyan, S., Sharma, D., Gangil, B., Patnaik, A., & Singh, T. (2020). Thermo-mechanical characterization of nonwoven fabric reinforced polymer composites. *Materials Today: Proceedings, 44*, 4770–4774. https://doi.org/10.1016/j.matpr.2020.10.972

Thakur, V. K., Singha, A. S., & Thakur, M. K. (2013). Ecofriendly biocomposites from natural fibers: Mechanical and weathering study. *International Journal of Polymer Analysis and Characterization, 18*(1), 64–72. https://doi.org/10.1080/1023666X.2013.747246

Väisänen, T., Das, O., & Tomppo, L. (2017). A review on new bio-based constituents for natural fiber-polymer composites. *Journal of Cleaner Production, 149*, 582–596. https://doi.org/10.1016/j.jclepro.2017.02.132

Van De Velde, K., & Kiekens, P. (2001). Thermoplastic pultrusion of natural fibre reinforced composites. *Composite Structures, 54*(2–3), 355–360. https://doi.org/10.1016/S0263-8223(01)00110-6

Verma, S. K., Gangil, B., Gupta, A., Rajput, N. S., & Singh, T. (2022). Dolomite dust filled glass fiber reinforced epoxy composite: Influence of fabrication techniques on physicomechanical and erosion wear properties. *Polymer Composites, 43*(1), 551–565. https://doi.org/10.1002/pc.26398

Verma, S. K., Gupta, A., Patel, V. K., Gangil, B., & Ranikoti, L. (2019). The potential of natural fibers for automotive sector. *Energy, Environment, and Sustainability*, 31–49. https://doi.org/10.1007/978-981-15-0434-1_3

Verma, S. K., Gupta, A., Singh, T., Gangil, B., Jánosi, E., & Fekete, G. (2019). Influence of dolomite on mechanical, physical and erosive wear properties of natural-synthetic fiber reinforced epoxy composites. *Materials Research Express, 6*(12). https://doi.org/10.1088/2053-1591/ab5abb

Zampaloni, M., Pourboghrat, F., Yankovich, S. A., Rodgers, B. N., Moore, J., Drzal, L. T., Mohanty, A. K., & Misra, M. (2007). Kenaf natural fiber reinforced polypropylene composites: A discussion on manufacturing problems and solutions. *Composites Part A: Applied Science and Manufacturing, 38*(6), 1569–1580. https://doi.org/10.1016/j.compositesa.2007.01.001

3 Effect of Processing on Natural Fibers for Composite Manufacturing

N. Vasugi, R. Sunitha, S. Amsamani, and C. Balaji Ayyanar

CONTENTS

3.1 INTRODUCTION

"Go Green" is the call from every part of the world. Man, who was a part of nature, has disturbed it to the maximum and today has created air, water and land pollution; therefore, man is forced to use the available resources and reuse it to create a sustainable environment. The attention towards the environment has led to renewal of natural and recyclable materials for its safety. Natural fibers are undergoing a high-tech revolution that could see them replace synthetic materials. The global requirement of eco-friendly technical textiles and its sustainability is proving the 'cradle-to-grave' concept is appreciable in the manufacturing of composite structures. In this concern bio composites and the green composites produced from natural sources are gaining importance. The major attractions about these composites are that they are environmentally friendly, fully degradable or recyclable and sustainable (Singh et al., 1996). These can be easily disposed of or composted without harming the environment.

DOI: 10.1201/9781003272625-3

43

They can be rightly called the next generation composites. The bio fibers serve as a reinforcement by enhancing the strength and stiffness of the resulting composite structures. The vital property of these composites can be listed as light weight, high strength and recyclability. The properties of natural fibers vary physically and in chemical constituents depending on the portion of plant from which the fibers are obtained, geographical conditions of the growth of the plant and age of the plant. The main advantage of the green and bio composites is the replacement of traditional materials such as steel Verma and Deepak (2016). Some natural fiber and natural resins commonly used in production of bio composites are hemp, flax, jute, kenaf and sisal fibers and are derived from starch, vegetable oils and protein. Today commercial products and applications are emerging for these green composites (Wang et al., 2015). The widely varied properties of these composites have increased their applications to components of vehicles from the simple bicycle to the high tech spaceship, mobiles and civil engineering works (Kandpal et al., 2015). The area even calls for in-depth researches across the manufacturing of value added products like anti-bacterial walls, sound absorbing panels, sunlight proof covers, fire retardancy, vibration damping, impact strength, gas barrier and waterproof composites. In the green composites, both matrix and fiber are completely biodegradable and are renewable (Thyavihalli et al., 2019). The bio composites are environmentally friendly in their different stages of production, processing and waste disposal. The chemical composition of different vegetable fibers and treatment of the fibers for modification of properties for composites have been reviewed.

3.2   NEED FOR ECO FRIENDLY COMPOSITES

The bio composites fall under the category of polymer matrix composites. Polymer matrix composites are made up of natural or synthetic matrix materials namely thermoplastic and thermosetting plastic with one or more reinforcements such as carbon fibers, glass fibers or natural fibers (Faruk et al., 2010 & Sharma et al., 2020). Synthetic matrix materials are polyethylene, polypropylene, polycarbonate, polyvinyl chloride, nylon, acrylics and carbon steel Kevlar, epoxy resins which are not biodegradable (Ku et al., 2011). The eco-friendly composites are advantageous due to the following reasons.

- Very simple production process
- Enhanced properties of the end product
- Reduced production cost due to low energy consumption
- Provide rural populations with a new source of income in economically deprived areas.

3.3   CLASSIFICATION OF SOURCES OF
ECO-FRIENDLY COMPOSITES

The eco-friendly composites, say bio or green composites, are divided into three main categories based on type of reinforcement and polymer materials, as totally renewable composites in which both matrix and reinforcement are from renewable resources, partly renewable composites in which matrix is obtained from renewable resources and reinforced from a synthetic material, and partly renewable composites in which synthetic matrix is reinforced from natural bio polymers. The classification can be made on the basis of the sources and their nature also. This has two main sub classes as wood and non-wood-based bio fibers. The wood-based fibers can be further classified as hard, soft and recycled wood fibers, where the non-wood fibers are subdivided as straw, bast, leaf, seed and grass type.

3.3.1   WOOD-BASED FIBERS

These are fibers obtained from the trees and plant stems. The wood may be hardwood or softwood. The hard fibers are those which are thick, stiff and coarse with irregular thickness and much less ability to be spun. The recycled wood fibers are ones which are modified by mechanical or chemical processes. Paper is a good example for products from wood fibers. Similarly, rayon, a cellulose fiber, is also obtained from wood pulp. This is a semi synthetic fiber.

3.3.2   NON-WOOD-BASED FIBERS

These fibers are obtained from various parts of a plant. The waste from rice and wheat after harvest is called straw fibers. Bast fibers are thread-like structures obtained from stems such as jute, hemp, okra, flax, abaca, kenaf, ramie and rattan; the leaf fibers are obtained from leaf portions of the plant such as sisal, curaua, henequen, agave and pineapple. The fruit fibers are from the fruit of the plants such as coir, oil and palm, and the seed fibers are seen on the seeds of the plant like cotton and kapok fibers which stick to the cotton seeds. The fibers obtained from the grass variety of plants are called grass fibers, for example bamboo fibers. These fibers are removed from the stem or leaf or husk by the process of retting or decortication (Ahmad et al., 2019).

3.4   NATURAL FIBERS AND COMPOSITES

The fibers obtained from plant sources are the most abundantly available fiber among all other natural fibers. Hemp, Himalayan nettle, sisal, jute, kenaf, flax, abaca and ramie are commonly known plant fibers. Plant fibers are also called cellulosic fibers and have quite promising tensile strength (Kocaman et al., 2017). Natural fibers like sisal, coir, hemp and oil palm are now finding applications in a wide range of industries. The research works carried out and ongoing in the field of bio fiber composites have experienced interest, particularly with regard to the comparable properties obtained in the natural fiber reinforced composites over the glass fiber reinforced composite structures. These bio composites find their applications in the automotive industry, predominantly in interior applications (Thwe & Liao, 2003). Although glass fibers have benefits such as low cost and high strength, these are non-biodegradable but the natural fibers are renewable, cheap, recyclable, biodegradable and also abundantly available. The natural fiber reinforced composites exhibit low density and reduce the extent of environmental pollution caused by synthetic fibers (George et al., 2001).

3.5   CHEMICAL CONSTITUENTS OF NATURAL FIBERS

The chemical properties of plant fibers are affected by the reasons such as climatic conditions, floral classification, stalk height, degradation, age and time taken by the plant while growing (Dittenber & GangaRao, 2011). The properties of natural fibers vary depending on the presence of chemical constituents which become responsible for imparting additional qualities to the composite structures Seena et al. (2005). The cellulose, hemicellulose, lignin and pectin are found in varying proportions in plant fibers. Some items are non-structural such as waxes, inorganic salts and nitrogenous substances (Nirmal et al., 2015). Chemical properties of natural fiber have their own significance and they form the basis of predicting the final properties of fiber polymer fabricated composite as they influence physical, mechanical, thermal and tribological properties when used as reinforcement (Bénard et al., 2007). Cellulose offers superior mechanical properties (Sajith et al., 2017). Hemicellulose is very hydrophilic, soluble in alkali and easily hydrolyzed in acids. Lignin is amorphous and hydrophobic in nature, a complex hydrocarbon polymer with both aliphatic and aromatic constituents. It reduces water absorption and enhances thermal stability (Mishra et al., 2001). Pectins is a collective name for heteropolysaccharides and they give plants flexibility. Bio fibers can be considered to be composites of hollow cellulose fibrils held together by a lignin and hemicellulose matrix (Jayaraman, 2003). The lumen is a hollow central cavity in a fiber cell, responsible for reducing the density, increasing thermal insulation and noise-resistance properties (Reddy & Yang, 2005). Utilization of certain types of lignocellulosic fibers exhibit higher mechanical properties and overall strength than that of synthetic fibers. (Vilaseca et al., 2010).

3.6   NATURAL FIBER PROPERTIES AND NEED
OF PROCESSING FOR COMPOSITES

The properties of the composite structures are influenced by the presence of various chemical constituents in the natural fibers when reinforced, and can be altered by fiber treatments. The thermal behavior of green composites fabricated from bagasse fiber and polyvinyl alcohol was observed to have an increase of thermal stability upon incorporation of bagasse fiber. Good permeability of composite structure is obtained by reinforcing sisal and flax fibers in resin (Fernandes et al., 2013). Kenaf fiber is a cellulosic fiber which has both economic and ecological advantages as it can be grown in a varied environment and the growth of the plant is very promising (Akil et al., 2011). Banana fiber is composed of mainly carbohydrates and protein which makes it a high strength fiber but it has content of water uptake. Thus, initial treatment is also needed for the banana fiber (Venkateshwaran & Elayaperumal, 2010). Areca fiber needs to be treated before using it for making composite to reduce the moisture absorption (Padmaraja et al., 2013). Likewise several other natural fibers have the inherent properties which may or may not require modification for making them suitable for composite preparation based on the end uses.

The main difficulties of natural fibers in composite material arises due to low percentage of fiber reinforcement by the weak strength. Naturally, the hydrophilic character of natural fibers tends to absorb more quantity of moisture, which often causes incompatibility with the polymer matrix (Jauhari et al., 2015). The chemical treatment enhances the properties of fiber by reducing the moisture content. The strongly polarized cellulosic fibers inherently are less compatible with hydrophobic polymers (Ranakoti et al., 2018).

The mechanical and physical properties of natural fibers vary significantly depending on the chemical and structural composition, fiber type and growth conditions. The essential mechanical properties of composite structures are highly influenced by the bonding between the matrix and fibers. The adhesion properties can be varied by pretreating the fibers. So, special processing methods such as physical and chemical modification methods are developed (Kocaman et al., 2017). The poor compatibility of natural fibers shows a major restriction on effective use of natural fibers in durable composites. Modifications of natural fiber have to be considered in transforming the fiber surface properties to improve their properties such as stability, high moisture absorption and adhesion with different matrices. The modifications of fiber are classified as physical and chemical (Franco & González, 2005).

3.7 TREATMENTS OF NATURAL FIBERS FOR COMPOSITES

Surface treatment of fibers improves the strength or bonding between matrix and fiber which leads to improved properties (Yegireddi et al., 2019). Generally, the fiber treatments are grouped as physical, chemical and biological treatments.

3.7.1 PHYSICAL TREATMENT

The physical treatment methods involve corona discharge, plasma, ionized air, ultraviolet radiation, electron radiation and thermal treatment (Ahmad et al., 2019). The structural properties of natural fibers are altered which results in increasing the mechanical bonding between fiber and polymer (Adekunle, 2015). The advantages of physical treatment are that these treatments improve the salient properties of the fibers thereby improving the properties of the end product, improve the compatibility between the fiber and matrix, have a low impact on the environment, consume less energy and use a low-cost process.

Surface modification by discharge methods namely plasma, sputtering and corona improve the functional properties of plant fibers. Low temperature plasma treatment causes mainly chemical implantation, etching, polymerization, free radical formation and crystallization, whereas sputter etching brings chiefly physical changes such as surface roughness and this leads to an increase in adhesion and decreases light reflection (Wakida & Tokino, 1996). Plasma treatment is a physical technique which has been successfully utilized to modify the surface of various natural fibers. Mechanical properties of natural fibers were found to improve significantly after plasma treatment (Oliveira et al., 2012). Plasma treatment introduces various functional groups on the natural fiber surface and these functional groups can form strong covalent bonds with the matrix leading to strong fiber/matrix interface. Also, surface etching due to plasma treatment may improve the surface roughness and result in better interface with the matrices through mechanical interlocking (Shahidi et al., 2013). Electrical discharge method improves the mechanical properties of composites. Corona treatment changes the surface energy of the cellulosic fibers, which in turn affects the melt viscosity of composites (Belgacem et al., 1994). Electron radiation improves interfacial bonding and reduces thermal decomposition (Ferreira et al., 2014; Zenkiewicz & Dzwonkowski, 2007). Ionized air increases the wetting property of fibers and interfacial enhancements of the composites is lower compared to chemical treatments (Razera & Frollini, 2004; Milanese et al., 2012). Thermal treatment improves the thermal stability and decreases the moisture content of the fibers. It also retains the chemical properties of the fibers (Manikandan et al., 2001; Manna et al., 2017). Ultraviolet radiation increases the wettability of the fibers and improves the mechanical properties of the composite structure. It is also cheap, flexible and easy to install, and hence it is widely attractive (Khan et al., 2004).

3.7.2 CHEMICAL TREATMENT

The modification of the natural fibers using chemicals alters the nature of plant fibers permanently by grafting polymers on the surface of the fibers. It may also happen with bulking or cross-linking within the fiber cell wall. The chemical modification provides more dimensional stability, reduces water absorption capacity and gives resistance to fiber against fungal decay (Xie et al., 2010). Chemical methods include some popularly known treatments such as graft copolymerization, silane treatments, alkali swelling, cyanate treatment and fiber impregnation which results in enhancement of the adhesion between the fiber and polymer (Adekunle, 2015). It is observed that some of these chemical treatments can significantly improve the mechanical properties of natural fibers by modifying their crystalline structure, as well as by removing weak components like hemicelluloses and lignin from the fiber structure (Shahidi et al., 2013). Chemical surface treatment has shown a sustainable mechanical performance of natural fiber composites when subjected to wet and humid conditions (Joseph et al., 1997). It also improves the adhesion between the fiber surface and polymer matrix by its capability of modifying nature in fibers.

On bleaching, the fiber becomes more uniform, acquires more swelling capacity and loses in tenacity and weight due to the removal of some increasing materials of cellulose (Farouqui & Hossain, 1989). Also, moisture absorption of natural fibers is reduced by appropriate chemical treatments. The chemical treatments with silane coupling agents can improve the fiber and matrix interfacial interactions through formation of strong chemical bonding which results in significant improvement in the mechanical properties of composites. Silanes used for treatment of fibers have different functional groups at either end such that interaction at one end can occur with hydrophilic groups of the fiber and the other end can interact with hydrophobic groups (Xie et al., 2010). Silanes used are amino, methacryl, alkyl silanes and glycidoxy which increase the hydrophobicity of natural fibers and strength of natural fiber composites (Rachini et al., 2012).

Alkaline treatment or mercerization is the chemical treatment done to natural fibers for composite manufacturing as this increases surface roughness. The aqueous sodium hydroxide (NaOH) is used to promote the ionization of the hydroxyl group to the alkoxide. The alkali treatment changes the color of the bleached fibers from white to pale yellow. Alkali imparts crimp in the fiber too (Pickering et al., 2003). Hemicellulose of the fiber is dissolved in concentrated alkali medium that causes weight loss in alkalization. Diameter of the fiber reduces more in thicker fibers and lesser in thinner fibers (Prasad et al., 1983). Among different methods, alkalization is an effective process to improve the fiber-matrix interaction, thermal stability and heat resistance (Zin et al., 2018). Surface treated fibers show higher thermal stability than untreated fiber (Amantes et al., 2017). The elimination of lignin and hemicellulose takes place due to mercerization, which results in increased cellulose content for treated fiber. The surface treated jute fiber reinforced composites hold higher degradation temperature resulting in improved thermal stability (Athiyamoorthy & Subramaniam, 2020). The moisture absorbed by the fibers can be reduced by chemical modifications of fibers such as acetylation, methylation, cyanoethylation, benzoylation, permanganate treatment and acrylation (Sreekala & Thomas, 2003). Benzoyl chloride is the most often used chemical for treating fibers. This increases the strength and decreases the water absorption thereby improving the fiber matrix bonding (Murali & Chandra, 2014). Acylation of natural fibers is a well-known esterification method which is divided into acetylation and valerylation causing plasticization of cellulosic fibers (Xie et al., 2010). Polymer hydroxyl groups of the cell wall with acetyl groups, modify the properties of these polymers so that they become hydrophobic which could stabilize the cell wall against moisture, improving dimensional stability and environmental degradation (Ku et al., 2011).

3.7.3 BIOLOGICAL TREATMENT

Enzyme treatment is a biological modification of fibers. It needs low energy input for the process, and it is environmentally friendly. This biological treatment is becoming an important technology due to its advantages in environmentally friendly impacts and economic wise (Murali et al., 2014). Enzymes act as accelerators in changing substrates into targeted products. An enzyme particle interacts with other particles in a substrate whose velocity of changes or activity depends on process conditions. Enzymes can be obtained from various microorganisms (Padzil et al., 2020). The deposition of bacterial cellulose nanofibrils on the natural fiber surface resulted in improvement in interfacial adhesion with polymeric matrices such as polylactic acid and cellulose acetate (Juliana & Raul, 2016).

Improvement of composite properties is done with the application of enzyme treatment which improves tensile and flexural strength of abaca/PP composites and increased surface area leading to increased interfacial bonding (Bledzki et al., 2010). The source used for biological treatment needs to be explored more for large scale production.

3.8 APPLICATION

Natural fibers have many advantages compared to synthetic fibers like low density, lower cost, acceptable specific properties and also, they are renewable and biodegradable (Sreekala et al., 2002). This has laid a strong foundation for natural fiber composites to emerge as realistic alternatives to replace the glass reinforced composites in many applications. The plant fibers such as banana, kapok, coir, sisal and jute have attracted the attention of researchers and engineers for application in aerospace, civil and automotive fields.

In the construction industry, wood fiber, polyethylene and polypropylene fibers are extensively used in decking. Plant fiber reinforced composites structures are gaining popularity in non-structural construction applications and also for door and window frames, wall insulation and floor lamination (Corradi et al., 2009). The eco-friendly fiber composites find their application in various fields and about 66% in packaging industries, 8% in textiles, 6% in agriculture, 4% in electronics, 4% in medical and 12% in other fields (Lalit et al., 2018).

3.9 CONCLUSION

Thus, the natural fibers vary in chemical constituents due to various solid reasons. These could be altered by the physical, chemical and biological treatments to make them suitable for various end uses. Physical treatments are disadvantageous as these are energy intensive. The chemical treatments can be easily carried out in controlled conditions but are not eco-friendly. Though biological treatment has its own advantages over chemical treatments, low scalability and difficulties in controlling processes become highly disadvantageous. Each and every method has its own advantages and disadvantages. So, an appropriate fiber treatment method should be utilized for treating the natural plant fibers. Eco friendly and recyclable composites in the names of green and bio composites find their potential in various fields of technical applications. These are lightweight and eco-friendly which may be used in build tech, mobile tech, home tech and many other applications. These may be cost effective too when these are manufactured on a large scale. Many more research works should be carried out in finding and fulfilling their applications in different fields.

REFERENCES

Adekunle, F. L. (2015). Surface treatments of natural fibres – A review: Part 1. *Journal of Polymer Materials*, *3*, 41–46.

Ahmad, R, Hamid, R., & Osman, S. (2019). Physical and chemical modifications of plant fibres for reinforcement in cementitious composites. *Advances in Civil Engineering*, *2019*, Article ID 5185806. https://doi.org/10.1155/2019/5185806.

Akil, H. M, Omar, M. F., Mazuki, A. A. M., Safiee, S, Ishak, Z. A. M., & Bakar, A. (2011). Kenaf fibre reinforced composites: A review. *Materials and Design*, 4107–4121.

Amantes, B. P., Melo, R. P., Neto, R. P. C., & Marques, M. F. V. (2017). Chemical treatment and modification of jute fiber surface. *Chemistry and Chemical Technology*, *11*, 333–343.54.

Athiyamoorthy, M. S., & Subramaniam, S. (2020). Mechanical, thermal, and water absorption behaviour of jute/carbon reinforced hybrid composites. *Sådhanå*, *45*, 278.

Belgacem, M. N., Bataille, P., & Sapieha, S. (1994). Effect of corona modification on the mechanical properties of polypropylene/cellulose composites. *Journal of Applied Polymer Science*, *53*(4), 379–385.

Bénard, Q., Fois, M., & Grisel, M. (2007). Roughness and fibre reinforcement effect on wettability of composite surfaces. *Applied Surface Science*, *10*, 4753–4758.

Bledzki, A. K., Mamun, A. A., Jaszkiewicz, A., & Erdmann, K. (2010). Polypropylene composites with enzyme modified abaca fiber. *Composites Science and Technology*, *70*(5), 854–860.

Corradi, S., Isidori, T. Corradi, M., & Soleri, F. (2009). Composite boat hulls with bamboo natural fibers. *International Journal of Materials and Product Technology, 36*(1), 73–89.

Dittenber, D. B., & GangaRao, H. V. S. (2011). Critical review of recent publications on use of natural composites in infrastructure. *Composites Part A, 43*(8), 1419–1429.

Farouqui, F. I., & Hossain, I. H. (1989). The Rajshahi University studies, Part-B.

Faruk, O., Bledzki, A. K., Fink, H.-P., & Sain, M. (2010). Biocomposites reinforced with natural fibers: 2000—Prog. *Polymer Sciences, 37*, 1552–1596.

Fernandes, E. M., Mano, J. F., & Reis, R. L. (2013). Hybrid cork—Polymer composites containing sisal fibre: Morphology, effect of the fibre treatment on the mechanical properties and tensile failure prediction. *Composite Structures, 105*, 153–162.

Ferreira, M. S., Sartori, M. N., Oliveira, R. R., Guven, O., & Moura, E. A. B. (2014). Short vegetal-fiber reinforced HDPE-A, study of electron-beam radiation treatment effects on mechanical and morphological properties. *Applied Surface Science, 310*, 325–330.

Franco, P. J. H., & González. A. V. (2005). Fiber-matrix adhesion in natural fiber composites. In *Natural Fibers, Plastics and Composites*. London: CRC Press Taylor & Francis Group.

George, J., Sreekala, M. S., & Thomas, S. (2001). A review on interface modification and characterization of natural fiber reinforced plastic composites. *Polymer Engineering and Science, 41*(9), 1471–1485.

Jauhari, N., Mishra, R., & Thakur, H. (2015). Natural fiber reinforced composite laminates—A review. *Materials Today: Proceedings, 2*, 2868–2877.

Jayaraman, K. (2003). Manufacturing sisal-polypropylene composites with minimum fibre degradation. *Composites Science and Technology, 63*(3–4), 367–374.

Joseph, K., Paul, A., & Thomas, S. (1997). Effect of surface treatments on the electrical properties of low-density polyethylene composites reinforced with short sisal fibres. *Composites Science and Technology, 57*, 67–79.

Juliana, C., & Raul, F. (2016). International Symposium on "Novel Structural Skins: Improving sustainability and efficiency through new structural textile materials and designs" Surface modification of natural fibers: A review 1877–7058.

Kandpal, B. C., Rakesh, C., & Vishal, K. (2015). Recent advances in green composites—A review. *International Journal for Technological Research in Engineering* (IJTRE), 2(7).

Khan, M. A., Shehrzade, S., & Hassan, M. M. (2004). Effect of alkali and ultraviolet (UV) radiation pretreatment on physical and mechanical properties of 1, 6- hexanediol diacrylate-grafted jute yarn by UV radiation. *Journal of Applied Polymer Science, 92*(1), 18–24.

Kocaman, S., Karaman, M., Gursoy, M., & Ahmetli, G. (2017). Chemical and plasma surface modification of lignocellulose coconut waste for the preparation of advanced bio based composite materials. *Carbohydrate Polymers, 159*, 48–57.

Ku, H., Wang, H., PattarachaiKoop, N., & Trada, M. (2011). A review on the tensile properties of natural fiber reinforced polymer composites. *Composites. Part B Engineering, 42*(4), 856–873.

Lalit, R., Mayank, P., & Ankur, K. (2018). Natural fibers and biopolymers characterization: A future potential composite material. *Journal of Mechanical Engineering—Strojnícky časopis, 68*(1), 33–50. https://doi.org/10.2478/scjme-2018–0004

Manikandan Nair, K. C., Thomas, S., & Groeninckx, G. (2001). Thermal and dynamic mechanical analysis of polystyrene, composites reinforced with short sisal fibres, *Composites Science and Technology, 61*(16), 2519–2529.

Manna, S., Saha, P., Chowdhury, S., Thomas, S., & Sharma, V. (2017). Alkali treatment to improve physical, mechanical and chemical properties of lignocellulosic natural fibers for use in various applications. *Lignocellulosic biomass production and industrial applications, 47*–63.

Milanese, A. C., Cioffi, M. O. H., & Voorwald, H. J. C. (2012). Thermal and mechanical behaviour of sisal/phenolic composites. *Composites Part B, Engineering, 43*(7), 2843–2850.

Mishra, S. B., & Luyt, A. S. (2008). Effect of organic peroxides on the morphological, thermal and tensile properties of EVA-organoclay nanocomposites. *Express Polymer Letters*, 2(4), 256–264.

Mishra, S. B., Misra, M., Tripathy, S. S., Nayak, S. K., & Mohanty, A. K. (2001). Graft copolymerization of acrylonitrile on chemically modified sisal fibres. *Macromolecular Materials and Engineering*, 286, 107–113.

Mollaert, M., & De Laet, L. (2017). Novel Structural Skins: Improving sustainability and efficiency through new structural textile materials and designs–COST Action. *Impact*, 2017(5), 46–48.

Murali, B., & Chandra Mohan, D. (2014). Chemical treatment on hemp/polymer composites *Journal of Chemical and Pharmaceutical Research*, 6(9), 419–423.

Nirmal, U., Hashim, J., & Megat Ahmad, M. M. H. (2015). A review on tribological performance of natural fibre polymeric composites. *Tribology International*, 83, 77–104.

Oliveira, F. R., Erkens, L., Fangueiro, R., & Souto, A. P. (2012). Surface modification of banana fibers by DBD plasma treatment, *Plasma Chemistry and Plasma Processing*, 32(2), 259–273.

Padmaraj, N. H., Kini, M. V., Pai, B. R., & Shenoy, B. S. (2013). Development of short areca fiber reinforced biodegradable composite material. *Procedia Engineering*, 64, 966–972.

Padzil, F. N. M., Ainun, Z. M. A. Kassim, A. N. Lee, N., S. H., & Lee, C. H. (2020). Pineapple leaf fibers. *Green Energy and Technology*. https://doi.org/10.1007/978-981-15-1416-6_5.

Pickering, K. I., Abdalla, A., Ji, C., McDonald, A. G., & Franich, R. A. (2003). The effect of silane coupling agents on radiata pine fiber for use in thermoplastic matrix composites. *Composites Part A*, 34(10), 1170–1178.

Prasad, S. V., Pavithran, C., & Rohatgi, P. K. (1983). Alkali treatment of coir fibres for coir-polyester composites. *Journal of Materials Science*, 18(5), 1443–1454.

Rachini, A., Troedce, M., Peyratout, C., & Smith, A. (2012). Chemical modification of hemp fibers by silane coupling agents. *Journal of Applied Polymer Science*, 123(1), 601–610.

Ranakoti, I. L., Pokhriyal, M., & Kumar. A. (2018). Natural fibers and biopolymers characterization: A future potential composite material. *Journal of Mechanical Engineering*, 68(1), 33–50.

Razera, I. A. T., & Frollini, E. (2004). Composites based on jute fibers and phenolic matrices: Properties of fibers and composites. *Journal of Applied Polymer Science*, 1077–1085.

Reddy, N., & Yang, Y. (2005). Biofibers from agricultural byproducts for industrial applications. *Trends Biotechnology*, 23(1), 22–27.

Sajith, S., Arumugam, V., & Dhakal, H. N. (2017). Comparison on mechanical properties of lignocellulosic flour epoxy composites prepared by using coconut shell, rice husk and teakwood as fillers. *Polymer Testing*, 58, 60–69.

Seena, J., Koshy, P., & Thomas, S. (2005). The role of interfacial interactions on the mechanical properties of banana fibre reinforced phenol formaldehyde composites. *Composite Interfaces*, 12, 581–600.

Shahidi, S., Wiener, J., & Ghoranneviss, M. (2013, January 16). *Surface modification methods for improving the dyeability of textile fabrics*. https://books.google.co.in/books?hl=en&lr=&id=IGOfDwAAQBAJ&oi=fnd&pg=PA33&dq=modification+methods+for+improving+the+dyeability+of+textile+fabrics&ots=zyDq4AOhDZ&sig=8YHQAV4qWYWZuUVmfmBBaof5z2Q&redir_esc=y#v=onepage&q=modification%20methods%20for%20improving%20the%20dyeability%20of%20textile%20fabrics&f=false

Sharma, A. K., Bhandari, R., Aherwar, A., & Rimašauskienė, R. (2020). Matrix materials used in composites: A comprehensive study. *Materials Today: Proceedings*, 21, 1559–1562.

Singh, B. Gupta, M., & Verma, A. (1996). Influence of fiber surface treatment on the properties of sisal—Polyester composites. *Polymer Composites*, 17(6), 910–918.

Sreekala, M. S., George, J., Kumaran, M. G., & Thomas, S. (2002). The mechanical Performance of hybrid phenol- formaldehyde—Based composites reinforced with glass and oil palm fibres. *Composites Science and Technology*, 62(3), 339–353.

Sreekala, M. S., & Thomas, S. (2003). Effect of fibre surface modification on water-sorption characteristics of oil palm fibres. *Composites Science and Technology*, *63*(6), 861–869.

Thwe, M. M., & Liao, K. (2003). Durability of bamboo-glass fiber reinforced polymer matrix hybrid composites. *Composites Science and Technology*, *63*(3–4), 375–387.

Thyavihalli Girijappa, Y. G., Mavinakere Rangappa, S., Parameswaranpillai, J., & Siengchin, S. (2019). Natural fibers as sustainable and renewable resource for development of eco-friendly composites: A comprehensive review. *Frontiers Materials*, *6*, 226.

Venkateshwaran, N., & Elayaperumal, A. (2010). Banana fibre reinforced polymer composites – A review. *Journal of Reinforced Plastics and Composites*, *29*(10), 2387–2396.

Verma, D., Gope, P. C., Zhang, X., Jain, S., & Dabral, R. (2016). Green composites and their properties: A brief introduction. In *Green approaches to biocomposite materials science and engineering* (pp. 148–164). IGI Global.

Vilaseca, F., Valadez-Gonzalez, A., Herrera-Franco, P. J., Àngels Pèlach, M., & Pere López, J. (2010). Biocomposites from abaca strands and polypropylene. Part I: Evaluation of the tensile properties. *Biosource Technology*, *101*.

Wakida, T., & Tokino, S. (1996). Surface modification of fibre and polymeric materials by discharge treatment and its application to textile processing. *Indian Journal of Fibre and Textile Research*, *21*, 69–78.

Wang, H., Wang, H., Schubel, P., Yi, X., Zhu, J., Ulven, C., & Qiu, Y. (2015). Green composite materials. *Advances in Materials Science and Engineering*, *2015*, 487416.

Xie, Y, Hill, C. A. S., Xiao, Z., Militz, H., & Mai, C. (2010). Silane coupling agents used for natural fiber/polymer composites: A review. *Composites Part A*, *41*(7), 806–819.

Xie, Y., Hill, C. A. S., Zefang, X., Militz, H., & Mai, C. (2010). Silane coupling agents used for natural fiber/polymer composites: A review. *Composites: Part A*, *4*(1), 806–819.

Xie, Y., Xiao, Z., Gruneberg, T., Militz, H., Hill, C. A. S., Steuernagel, L., & Mai, C. (2010). Effects of chemical modification of wood particles with glutaraldehyde and 1,3-dimethyylol-4,5-dihydroxyethyleneurea on properties of the resulting polypropylene composites. *Composites Science and Technology*, *70*(13), 2003–2011.

Yegireddi, S., Govind, N., & Kiran Kumar, C. (2019). *International Journal of Scientific and Technology Research*, *8*(8). www.ijstr.org.

Zenkiewicz, M., & Dzwonkowski, J. (2007). Effects of electron radiation and compatibilizers on impact strength of composites of recycled polymers. *Polymer Testing*, *26*(7), 903–907.

Zin, M. H., Abdan, K, Mazlan, N., Zainudin, E. S., & Liew, K. E. (2018). The effects of alkali treatment on themechanical and chemical properties of pineapple leaf fibres (PALF) and adhesion to epoxy resin. *IOP Conference Series: Material Science and Engineering*, *368*, 012035.

4 Mechanical Characterization of Labeo Catla and Laevistrombus Canarium Derived Hydroxyapatite-High Density Polyethylene Composite

C. Balaji Ayyanar, K. Marimuthu, C. Bharathira,
B. Gayathri, and S. K. Pradeep Mohan

CONTENTS

4.1 INTRODUCTION

Rice Husk Fiber (RHF), Bagasse Fiber (BF), and Waste Fish (WF) are used as reinforcing biodegradable agents for thermoplastic composites. In general, the addition of RHF and BF promoted an increase in the mechanical property when compared to the neat Poly Propylene (PP). This considerably improved the biodegradation of the composites and led to a higher degradation rate (Nourbakhsh et al., 2014). The properties of kenaf fiber (3 mm length, 134.3 kg/m³), HDPE (density of 961 kg/m³ and melt flow

DOI: 10.1201/9781003272625-4

index of 7 g/10 min) composites processed by extrusion were investigated for low and high processing temperature applications. Composites processed at high processing temperatures (HPT) showed better tensile strength compared to the ones at low processing temperatures (LPT) (Salleh et al., 2014). The effects of Azodicarbonamide (AZD) and Nanoclay (NC) content on the physio-mechanical and foaming properties of HDPE/Wheat Straw Flour (WSF) composites tensile strength gradually decreased from 20.23 MPa to 16.11 MPa with an increase in AZD from 2 phr to 4 phr (per hundred parts of resin). But by adding the NC from 0 to 2 phr, the composite tensile strength increased dramatically (Babaei et al., 2014).

The phosphate (alkaline) and calcium release were noticed in bone alternates which were prepared with Terminalia Arjuna (TA) extract when compared to control. The cell line MG 63 exhibited potential influence in cell differentiation (Krithiga et al., 2014, Aji et al., 2001; Alok & Alok, 2017). Many materials have been used for bone defect repair, but there is no ideal and fully approved implant. Among different materials, HDPE based composites can be used as implants, because of their high mechanical properties (E-modulus, strength, and hardness) and good biological properties (no toxicity and biocompatibility) (Pourdanesh et al., 2014). The property of shells filled with silane treated precipitated calcium carbonate and wood fiber core-shell structured wood plastic composites were studied. The flexural strength of the strong core (34.5 MPa) was greater than that of the weak core (14.9 MPa) (Kim et al., 2013). Composites such as AISI 316L, CoCrMo, Ti6Al4V, ultra-high-molecular-weight polyethylene, alumina, zirconia, polyether ether ketone, hydroxyapatite (48 MPa), and polytetrafluoroethylene are used as a biomaterials (Affatato et al., 2015). The combination of peat ash blended to HDPE composites was characterized. It was found that tensile strengths (33.52 MPa) and flexural modulus (1370.02 MPa) were higher than virgin HDPE (29.82 MPa).

The recycled paper blended in HDPE composite MFI of 7 g/10 min (190 °C/2.16 kg), a tensile strength of 17 MPa was fabricated by traditional injection molding or by turbo mixing. The best fibers dispersion was obtained by turbo mixing and the composites produced in this way have been added with maleated polyethylene in the amounts of 1, 3, and 5 wt. % (Valente et al., 2016). Though, after cooling the composite at 2 °C min⁻¹, the HDPE melting temperature reduced from 141 °C for the neat polymer to 136 °C (Dikobe & Luyt, 2017; Al-Oqla & Salit, 2017; Amir et al., 2014; Balaji Ayyanar & Marimuthu, 2019; Balaji Ayyanar et al., 2020). Banana particulates reinforced polyvinyl chloride was molded using compression molding with lightweight, low-cost materials, and good mechanical properties (UTS of 42 MPa) (Danasabe, 2017; Bashar, 2017). Though, after cooling the composite at 2 °C min⁻¹, the HDPE melting temperature reduced from 141 °C for the neat polymer to 136 °C (Dikobe, 2017; Birm-June et al. 2013). The waste bio fiber particulate blended with HDPE can be used as a substitute material for existing fastener material (Visakh et al., 2012).

Polymer membranes designed using polymer blends offer myriad opportunities as different functional groups can be harnessed on the surface that can render antibacterial/antifouling surface (Mural et al., 2018). Thermal investigation data showed crystallinity for HDPE/mica (size of 17 µm, an aspect ratio of 1.7) increasing with increasing filler content, but HDPE/wollastonite (size of 49 µm, an aspect ratio of 13.5) composites showed higher amorphous polymer phase substance (Lapčík et al., 2018). SSFs (Stainless Steel Flakes) blended HDPE composites while increasing the SSFs showed only small-scale changes of Ultimate Tensile Strength (UTS). The formation of bigger SSFs aggregates leads to stress accumulation and thereby a drop in tensile strength (Seretis et al., 2018; Fauzani Md, 2014, Fereydoun et al., 2014, Ishagh et al., 2014).

4.2 COMPRESSION MOLDING

The compression molding is one of the methods of the molding in which the required material is generally preheated and it is initially kept in a heated mold cavity. Different weight percentages 0, 10, 20, 30, 40, and 50 wt. % of (i) fish scale, (ii) seashell, and (iii) combinations of fish scale seashell particulates filled with HDPE composites were molded in compression molding with a size of $3 \times 150 \times 150$ mm (Lubomir et al. 2018; Maiju et al., 2013; Marco et al., 2016; Mehdi et al., 2010).

4.3 PREPARATION OF SPECIMENS

The molded specimens were prepared as per ASTM standards. Different tests like tensile strength (ASTM D638), compression strength (ASTM D695), and flexural strength (ASTM D790).

4.4 MECHANICAL CHARACTERIZATION

4.4.1 TENSILE STRENGTH

The different weight percentages (0, 10, 20, 30, 40, and 50 wt. %) of (i) fish scale, (ii) seashell, and (iii) combinations of fish scale and seashell particulates filled HDPE composite specimens were carried out tensile strength as per the standard ASTM 638. The tensile strength is gradually increased by varying the particulate contents from 0 wt. % up to 30 wt. %. The strength gradually decreased by increasing the particulates content of more than 30 wt. % which was observed in the test. Among the combinations the maximum tensile strength of the 30 wt. % of fish scale particulates filled HDPE composite specimen was found as 29.5 ± 0.5 MPa (Sahar et al., 2017).

4.4.2 COMPRESSIVE STRENGTH

The different weight percentages (0, 10, 20, 30, 40, and 50 wt. %) of (i) fish scale, (ii) seashell, and (iii) combinations of fish scale and seashell particulates filled HDPE composite specimens were carried out compressive strength as per the standard ASTM D695. The compressive strength of (i) fish scale, (ii) seashell, and (iii) combinations of fish scale and seashell particulates filled HDPE composite gradually increased by increasing the natural particulates reinforcement. The compressive strength gradually improved from 62 MPa to 74 MPa for the 0 wt. % to 50 wt. % of fish scale and seashell particulates filled HDPE composites respectively (Saverio et al., 2015).

4.4.3 SHORE D HARDNESS TEST

The different weight percentages (0, 10, 20, 30, 40, and 50 wt. %) of (i) fish scale, (ii) seashell, and (iii) combinations of fish scale and seashell particulates filled HDPE composite specimens were carried out shore D hardness test as per the ASTM D2240 standard for each specimen. The shore D hardness of the (i) fish scale, (ii) seashell, and (iii) combinations of fish scale and seashell particulates filled HDPE composite was gradually increased by increasing the natural particulates reinforcement. The specimens without reinforcement hardness were found 55 SHN and maximum shore D hardness 69 SHN were found for 50 wt. % combinations of fish scale and seashell particulates filled HDPE composites, respectively.

4.4.4 FLEXURAL STRENGTH

The different weight percentages (0, 10, 20, 30, 40, and 50 wt. %) of (i) fish scale, (ii) seashell, and (iii) combinations of fish scale and seashell particulates filled HDPE composite specimens were carried out the flexural test as per the ASTM D790 standard for each specimen. The flexural strength of the composite increased from 0 wt. % up to 30 wt. % of reinforcement; after that it gets decreased. The maximum flexural strength of the 30 wt. % combinations of fish scale and seashell particulates filled HDPE composite specimens are found as 36 ± 0.5 MPa, respectively.

4.5 CONCLUSIONS

The different weight percentages (0, 10, 20, 30, 40, 50 wt. %) of the composite were molded using the compression molding method and mechanical characterization was carried out. Among the combinations the 30 wt. % fish scale filled HDPE composite exhibited the highest tensile strength of 29.50 ± 0.5. The highest compressive strength was found to be 74 MPa for the 50 wt. % of the combinations of fish scale and seashell particulates filled HDPE composite. The maximum shore D hardness was found 69 for 50 wt. % of fish scale and seashell particulates filled HDPE composite. The maximum flexural strength of 30 wt. % combinations of fish scale and seashell particulates filled HDPE composite specimens were found as 36 ± 0.5 MPa.

REFERENCES

Affatato, S., Ruggiero, A., & Merola, M. (2015). Advanced biomaterials in hip joint arthroplasty. A review on polymer and ceramics composites as alternative bearings. *Composites Part B: Engineering, 83*, 276–283.

Aji, P. M., Packirisamy, S., & Sabu, T. (2001). *Studies on the Thermal Stability of Natural Rubber/Polystyrene Interpenetrating Polymer Networks: Thermogravimetric Analysis. Polymer Degradation and Stability*. London: Elsevier.

Alok, A., & Alok, S. (2017). *Mechanical, Thermal and Dielectric Behavior of Hybrid Filler Polypropylene Composites*. London: Elsevier.

Al-Oqla, F. M., & Salit, M. S. (2017). *Materials selection for natural fiber*

Amir, N., Alireza, A., & Ali, K. T. (2014). *Characterization and Biodegradability of Polypropylene Composites Using Agricultural Residues and Waste Fish*. London: Elsevier.

Babaei, I., Madanipour, M., Farsi, M., & Farajpoor, A. (2014). Physical and mechanical properties of foamed HDPE/wheat straw flour/nanoclay hybrid composite. *Composites Part B: Engineering, 56*, 163–170.

Balaji Ayyanar, C., & Marimuthu, K. (2019). *Investigation on the Morphology, Thermal Properties, and In-vitro Cytotoxicity of the Fish Scale Particulates Filled High-density Polyethylene Composite.* New York: Sage Publications.

Balaji Ayyanar, C., Marimuthu, K., Gayathri, B., & Sankarrajan. (2020). *Characterization and In vitro Cytotoxicity Evaluation of Fish Scale and Seashell Derived Nano-hydroxyapatite High-density Polyethylene Composite.* New York: Sage Publications

Bashar, D. (2017). *Thermo-mechanical Characterization of Banana Particulate Reinforced PVC Composite as Piping Material.* London: Elsevier.

Birm-June, K., Fei, Y., Guangping, H., Qingwen, W., & Qinglin, W. (2013). *Mechanical and Physical Properties of Core-shell Structured Wood-plastic Composites: Effect of Shells with Hybrid Mineral and Wood Fillers.* London: Elsevier.

Dikobe, D. G., & Luyt, A. S. (2017). *Thermal and Mechanical Properties of PP/HDPE/Wood Powder and MAPP/HDPE/Wood Powder Polymer Blend Composites.* London: Elsevier.

Fauzani Md, S., Aziz, H., Rosiyah, Y., & Ahmad Danial, A. (2014). *Effects of Extrusion Temperature on the Rheological, Dynamic Mechanical and Tensile Properties of Kenaf Fiber/HDPE Composites.* London: Elsevier.

Fereydoun, P., Ali, J., Seyedhossein, H., & Azra, A. (2014). *In-vitro and In-vivo Evaluation of a New Nanocomposite, Containing High-Density Polyethylene, Tricalcium Phosphate, Hydroxyapatite, and Magnesium Oxide Nanoparticles.* London: Elsevier.

Ishagh, B., Mostafa, M., Mohammad, F., & Arash, F. (2014). *Physical and Mechanical Properties of Foamed HDPE/Wheat Straw Flour/Nano Clay Hybrid Composite.* London: Elsevier.

Kim, B. J., Yao, F., Han, G., Wang, Q., & Wu, Q. (2013). Mechanical and physical properties of core–shell structured wood plastic composites: Effect of shells with hybrid mineral and wood fillers. *Composites Part B: Engineering, 45*(1), 1040–1048.

Krithiga, G., Hemalatha, T., Deepachitra, R., Kausik, G., & Sastry, T. P. (2014). *Study on Osteopotential Activity of Terminalia Arjuna Bark Extract Incorporated Bone Substitute.* Bengaluru: Indian Academy of Sciences.

Lapčík, L., Maňas, D., Lapčíková, B., Vašina, M., Staněk, M., Čépe, K., . . ., & Rowson, N. A. (2018). Effect of filler particle shape on plastic-elastic mechanical behavior of high density poly (ethylene)/mica and poly (ethylene)/wollastonite composites. *Composites Part B: Engineering, 141*, 92–99.

Lubomir, L., David, M., Barbora, L., Martin, V., Michal, S., Klara, C., Jakub, V., Kristian, E. W., Richard, W. G., & Neil, A. R. (2018). *Effect of Filler Particle Shape on the Plastic-elastic Mechanical Behavior of High-density Poly(ethylene)/mica and Poly(ethylene)/wollastonite Composites.* London: Elsevier.

Maiju, H., Aji, P. M., & Kristiina, O. (2013). *Bionanocomposites of Thermoplastic Starch and Cellulose Nanofibers Manufactured Using Twin-screw Extrusion.* London: Elsevier.

Marco, V., Jacopo, T., Alessia, Q., & Carlo, S. (2016). *Use of Recycled Milled-paper in HDPE Matrix Composites.* London: Elsevier.

Mehdi, J., Jalaluddin, H., Aji, P. M., & Kristiina, O. (2010). *Mechanical Properties of Cellulose Nanofiber (CNF) Reinforced Polylactic Acid (PLA) Prepared by Twin-screw Extrusion.* London: Elsevier.

Mural, P. K. S., Madras, G., & Bose, S. (2018). *Polymeric Membranes Derived from Immiscible Blends with Hierarchical Porous Structures, Tailored Bio-interfaces and Enhanced Flux: Potential and Key Challenges.* London: Elsevier.

Nourbakhsh, A., Ashori, A., & Tabrizi, A. K. (2014). Characterization and biodegradability of polypropylene composites using agricultural residues and waste fish. *Composites Part B: Engineering, 56*, 279–283.

Pourdanesh, F., Jebali, A., Hekmatimoghaddam, S., & Allaveisie, A. (2014). In vitro and in vivo evaluation of a new nanocomposite, containing high density polyethylene, tricalcium phosphate, hydroxyapatite, and magnesium oxide nanoparticles. *Materials Science and Engineering: C*, *40*, 382–388.

Sahar, S., Gilberto, S., Tanja, Z., & Aji, P. M. (2017). *3D Printing of Nano-cellulosic Biomaterials for Medical*. New York: Springer.

Salleh, F. M., Hassan, A., Yahya, R., & Azzahari, A. D. (2014). Effects of extrusion temperature on the rheological, dynamic mechanical and tensile properties of kenaf fiber/HDPE composites. *Composites Part B: Engineering*, *58*, 259–266.

Saverio, A., Alessandro, R., & Massimiliano, M. (2015). *Advanced Biomaterials in Hip Joint Arthroplasty, A Review on Polymer and Ceramics Composites as Alternative Bearings*. London: Elsevier.

Seretis, G. V, Manolakos, D. E., & Provatidis, C. G. (2018). *On the Stainless Steel Flakes Reinforcement of Polymer Matrix Particulate Composites*. London: Elsevier.

Valente, M., Tirillò, J., Quitadamo, A., & Santulli, C. (2017). Paper fiber filled polymer. Mechanical evaluation and interfaces modification. *Composites Part B: Engineering*, *110*, 520–529.

Visakh, P. M., Sabu, T., Kristiina, O., & Aji, P. M. (2012). *Natural Rubber Nanocomposites Reinforced with Cellulose Whiskers Isolated from Bamboo Waste: Processing and Mechanical/ Thermal Properties*. London: Elsevier.

5 Fabrication and Characterization of Lead-Free Magnetic-Ferroelectric Green Composites for Spintronic Applications

Manish Kumar, Subhash Sharma, Arvind Kumar, and Abhay Nanda Srivastava

CONTENTS

5.1 INTRODUCTION

Coexistence of ferromagnetic (FM) and ferroelectric (FE) properties in the same material, confirm the multiferroic nature of that particular material and coupling in the both order parameters thereby confirming the evidence of magnetoelectric coupling (Matsukura et al., 2015). In this connection, the composites of FM-FE in the lead-free preparation have centre of attention of the current researchers due to the high demand in various device applications (Fiebig & Trassin, 2016; Tokura et al., 2017; Spaldin & Ramesh, 2019).

A lot of research has been completed and is underway on the multiferroics. But the search for good multiferroic materials with required properties such as room temperature multiferroism, best coupling in the magnetic and ferroelectric parameters are still

DOI: 10.1201/9781003272625-5

an open area for the use in the various device application such as spintronic devices. So, keeping in mind of these requirements, efforts are required to explore the multiferroic field for the room temperature and highly stable multiferroic materials. Although, the coupling in between FE and FM parameters is not an easy task because of their different chemistry and very famous problem "$d^0 vs d^n$". The origin of ferroelectric property is based on fully filled outermost orbit or d0-ness, and the origin behind the ferromagnetic property is partially filled outermost orbit (Spaldin & Ramesh, 2019).

Further, there are a number of reports available in the current literature, which provides the information about the multiferroism in the various FM-FE lead-free composites. Chauhan et al. reported on the $CoFe_2O_4$-$BaTiO_3$ (CFO-BT) lead-free composites, prepared via solid state reaction technique. In this study, they have reported the improvement in some very useful properties such as structural, electric, magnetic, dielectric and magnetoelectric coupling. They have outstandingly achieved the magnetoelectric coupling coefficient value ~165.32μV/cm. Oe, which is very useful in the spintronic device applications (Mo et al., 2021). In some other important studies, Bangruwa et al. (2018) have investigated the lead-free multiferroic nanocomposites of $Bi.9Pr.1FeO3$-$Ni_5Zn_5Fe_2O_4$ (BPFO-NZFO) prepared via the sol-gel route and reported the anomalous magnetic-ferroelectric behaviour of the composite materials. With these results, with the addition of NZFO into the BPFO material, the multiferroic nature of BPFO enhances and makes it useful for the multiferroic material based applications (Shankar et al. 2019). In some of the studies by our group in various publications (Kumar et al., Shankar et al. and Dabas et al.) investigated the $LaSrMnO_3$-$BaTiO_3$ (LSMO-BT), $BaTiO_3$-$CoFe_2O_4$ (BT-CFO), CFO-BT and $BiFeO_3$-$CoFe_2O_4$ (BFO-CFO) lead-free composites via solid state reaction techniques. We have confirmed enhancement in the ferroelectric, ferromagnetic with multiferroic properties by these various composites and explained the use of these materials in various device applications such as memory, energy storage and other multiferroic-magnetoelectric materials based devices (Kumar & Ghosh, 2015; Sharma et al., 2014).

So, in this way, we have analyzed the results present in the latest literature on the various lead-free composites. The motive behind this chapter is very clear of investigation of the ferromagnetic, ferroelectric and multiferroic properties of lead-free magnetic-ferroelectric composites for the spintronic device applications.

5.2 CHARACTERIZATION TECHNIQUES

5.2.1 STRUCTURAL PROPERTIES

The structural investigation of any composite is the primary requirement for the phase formation. Powder X-ray diffraction (XRD) is one of the best tools to investigate the phase of the composite materials. In the line of magnetic-ferroelectric composites, there are a number of XRD reports available for better understanding. For instance, recently, Chauhan et al. have discussed the phase formation of $CoFe_2O_4$-$BaTiO_3$ (CFO-BT) multiferroic composites. The tetragonal symmetry for CFO and cubic symmetry for BT ferrites were confirmed individually and variation (increase) in the composite peaks is noted with the increase in CFO content into the CFO-BT composites and the data were refined with Rietveld refinement (Chauhan & Agarwal, 2021). In

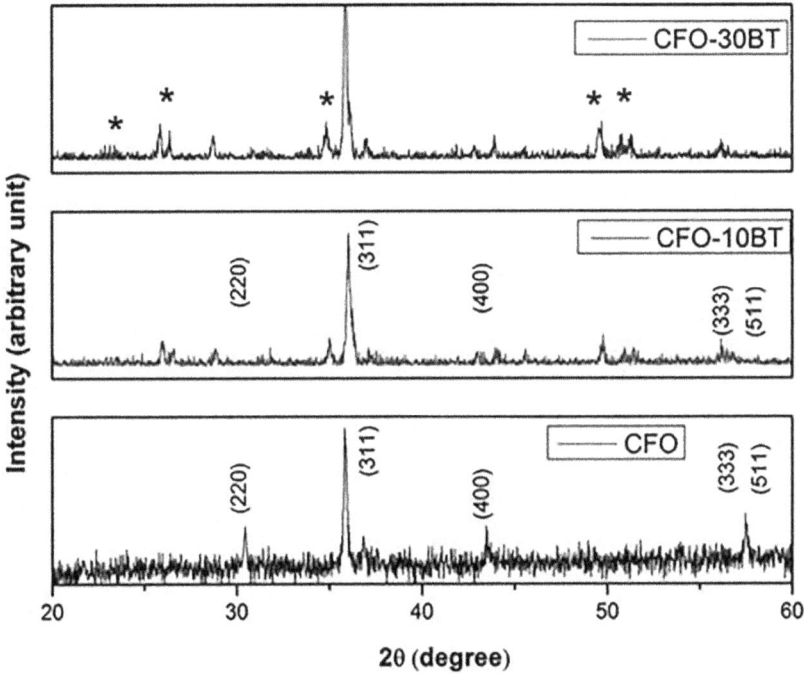

FIGURE 5.1 XRD patterns of magnetic (CFO)-ferroelectric (BT) composites.

Source: Shankar et al. (2019)

another study by our group, Shankar et al. discussed the CFO-BT composites formation via solid state reaction technique and confirmed the phase via standard JCPDS #22–1086 with space group Fd3m, as shown in Figure 5.1. The authors have confirmed the distortion in the CFO structure with the addition of BT, which may be due to the different ionic radii of Co^{2+} and Ba^{2+} and verified in the shifting of the peaks of the XRD spectra (Shankar et al., 2019).

5.2.2 Multiferroic Properties

Magnetization M vs H measurements at room temperature are explored in the various investigations and confirm the magnetic nature of the magnetic-ferroelectric composites (Chauhan & Agarwal, 2021; Bangruwa et al., 2018; Kumar & Ghosh, 2015; Shankar et al., 2018, 2019). Shankar et al. in CFO-BT composites investigated and measured the M vs H loops under the applied magnetic field of 15000 Oe at room temperature as shown in Figure 5.2(a). It is verified that the magnetization M vs H attains the saturation magnetization at higher field and the saturation magnetization decreases with the increase of the ferroelectric component into the CFO-BT composites. Authors have also confirmed that these materials may be used in the memory applications due to the enhancement in the remanency of the materials. The coercivity is the main part of these

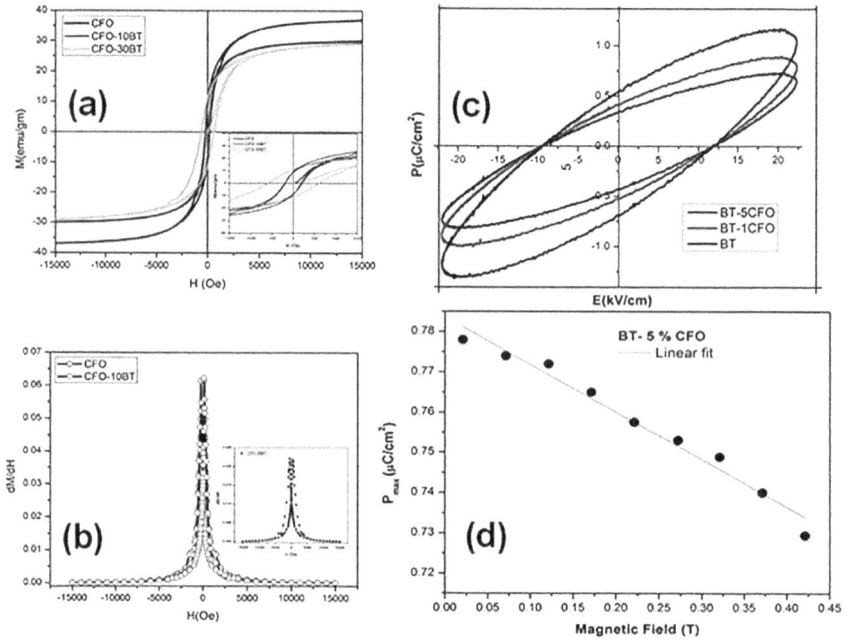

FIGURE 5.2   (a) M vs H hysteresis loops of pure CFO and CFO-BT composites. Inset shows the enlarged view of pure CFO and CFO-BT composites. (b) dM/dH vs H curve for pure CFO and CFO-BT composites for the confirmation of coercivity. (c) Ferroelectric hysteresis loops (P vs E) of pure BT and BT-CFO composites. (d) Maximum polarization of BT–5%CFO composite with the variation of applied magnetic field and straight line fit graph of BT–5%CFO composite.

Source: Reproduced with permission from Shankar et al. (2019).

M vs H loops for the various memory based applications so authors have included it in the enlarged view of M vs H loops in the inset (Figure 5.2(a)) of the figures and re-plotted as dM/dHvs H plots for the confirmation (Figure 5.2(b)) (Shankar et al., 2019).

The multiferroism i.e., simultaneous presence of ferroelectric as well as magnetic properties in the materials may be verified via the ferromagnetic hysteresis loop M vs H and ferroelectric hysteresis loop P vs E measurements. There are number of reports available in the literature for the P vs E loop (Chauhan & Agarwal, 2021; Bangruwa et al., 2018; Shankar et al., 2018, 2019; Dabas et al., 2019). It is reported by Shankar et al. that at some fixed frequency of 50 Hz, BT-CFO composites depict the extensive ferroelectric nature after applying 25kV electric field as shown in Figure 5.2(c). The ferroelectric nature is continuously decreasing in trend with the increase of the magnetic materials CFO into the ferroelectric BT (Shankar et al., 2018).

Another important measurement is reported for the multiferroic and magnetoelectric nature in the BT-CFO composites by Shankar et al. (2018). For the confirmation of the multiferroic as well as magnetoelectric nature in these composites, they have investigated the P vs E loop measurements under the presence of variable applied mag-netics and noted the maximum change in the maximum polarization values, as shown in Figure 5.2(d). They have plotted the maximum changed polarization values with respect to the applied magnetic field and noted the change in the maximum polarization values with the change in the applied magnetic field values. It confirms the evidence of multiferroism as well as magnetoelectric coupling in these samples (Shankar et al., 2018).

5.2.3 MAGNETOELECTRIC COUPLING

In materials science, the term multiferroic has attracted much attention in the past few years. The multiferroic and composite materials associate two or more ferroic properties, like FE and FM. The ability to control magnetic order by the means of an electric field or electric polarization (P) through a magnetic field opens a new door to technological innovation in the fields of information storage, sensing and computing (Jain et al., 2015; Béa et al., 2006; Ederer & Spaldin, 2005). Likewise, the Oersted field currently used to change the magnetic state of a commercial magnetic tunnel extends spatially and needs a moderate current to be generated. Such properties limit the areal density of the magnetic tunnel junction (Lee et al., 2008; Béa et al., 2005). Because the electrostatic field can be inadequate and desires very little current to produce, the incorporation of multiferroic materials (systems with dissimilar components coupling M and P order parameters) in a magnetic junction can enable a single memory solu-tion. The nonvolatile memory measured extra effective energy, faster, large in capacity and more reasonable than challenging technologies.

Strong magnetoelectric (ME) coupling allows a wide range of applications, consid-ering actuators, logic devices, ME sensors and non-volatile random-access memory (RAM). When compared with single phase ME materials, ME compounds composed of ferroelectric phase and ferromagnetic phase have attracted attention due to their rel-atively high ME coefficient and higher operating temperature. Advances in material synthesis technology have promoted the research on ME compounds, including vertical hetero-epitaxial nanocomposites (called 1–3 structures), multilayer thin films (called 2–2) and heterostructure nano-dots (called 0–0) (Lee et al., 2021; Ghidini et al., 2015; Okabayashi et al., 2019; Lu et al., 2011; Tian et al., 2016; Li et al., 2015). In the past, many BFO-based composite materials have been studied for improved magnetoelectric coupling, such as BFO-CFO, BFO-LSMO, BTO-LSMO, BFO-PZT (Jain et al., 2015; Mo et al., 2021; Yu et al., 2012), but in this chapter, we will only focus on lead-free materials. Among these, self-assembled perovskite columnar epitaxial spinel nanocom-posites, such as $BiFeO_3$-$CoFe_2O_4$ (BFO-CFO), have been extensively studied. Such nanocomposites show large epitaxial tension at the vertical spinel perovskite interface, allowing relatively high ME coupling and tunable electronic, magnetic and multi-ferroic properties. As a well-known research material, $BiFeO_3$ (BFO) is considered a single-phase multiferroic material that exhibits magnetoelectric coupling between antiferromagnetic and ferroelectric parameters at a temperature hundreds of degrees higher than the ambient temperature. Therefore, BFO may be an attractive technical material. Figure 5.3(a, b) demonstrate the basics of magnetoelectric couplings and dif-ferent kinds of couplings which is possible in different materials (Wang et al., 2010).

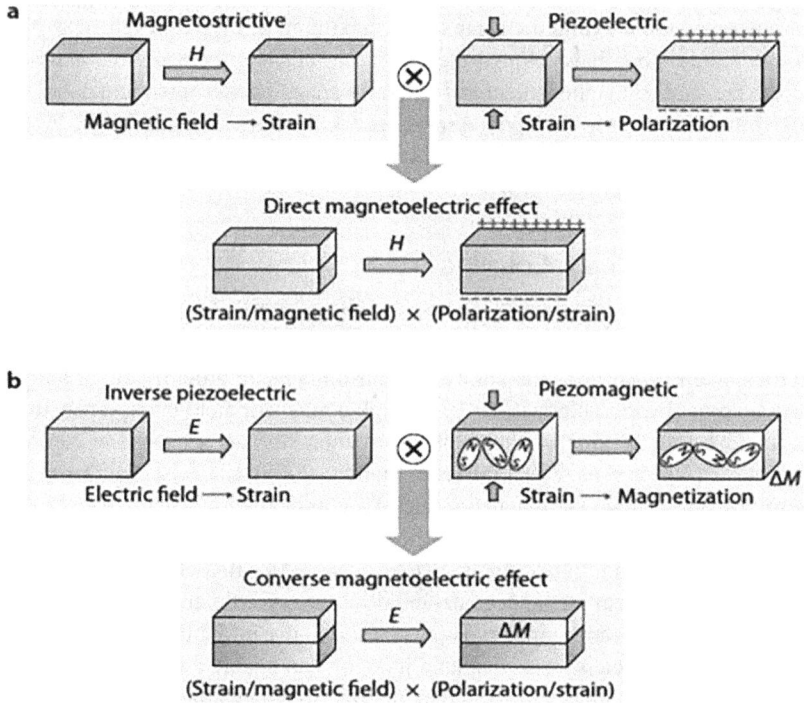

FIGURE 5.3   Graphic representation of strain-mediated ME coupling effect in composites containing FM layer (purple) and FE layer (pink), (a) direct ME effect, (b) reverse ME effect.

Source: Reproduced with permission from Wang et al. (2010)

However, significant challenges prevent progress. As we know, the electrical polarization vector may be along any of the possible eight equivalent directions, therefore the state of the polarization domain is badly defined. The next is, the magnetization of the subnetwork which has six easy state of the antiferromagnetic domain and is poorly defined in spite of the polarization which was saturated. Another well-known point is that the BFO is an antiferromagnetic which means there is no net magnetization in the bulk part. The theoretical touch given by Ederer and Spaldin proposes that the inclination of the antiferromagnetic structure due to the Dzyaloshinskii-Moriya interaction (DM) could make a small uncompensated moment of 0.1 mB/Fe (0.15kA/m) in BFO—corresponding to a magnetization approximately two orders smaller than that of a ferromaiman in a tunnel junction. The DM interaction is improbable to be a beneficial magnetic handle.

5.3   LATEST APPLICATIONS IN SPINTRONIC DEVICES

Among all the different BFO-based heterostructures, we will discuss here the magnetoelectric coupling in the interface between LSMO and BFO using spins to control the magnetism and which further are very important in spintronics devices. Due to the reconstruction of the global electronic orbital in the hetero-interface, the new magnetic state has been emerged. This can further cause the exchange bias to be coupled through a heterogeneous interface. Using the coupling between the rotational degrees of freedom and the orbital degrees of freedom, you can also find new ways to use electric fields to control the magnetism of the system. It is well known that the degree of freedom of the orbit is linked with occupancy rate of the d electron. In principle, the modification in the electronic state of the interface caused by the modification in ferroelectric polarization direction can modulate the magnetic coupling of the interface, and finally lead to the magnetoelectric coupling.

In order to verify this, a typical ferroelectric field effect transistor device was fabricated (Ramesh & Spaldin, 2007) as shown in Figure 5.4(a), where a BFO layer was used as a ferroelectric gate to control the charge state in the LSMO channel interface. For the rising (falling) ferroelectric polarization state, holes will accumulate (deplete) in the interface through the ferroelectric shield. Therefore, the sheet resistance between different ferroelectric polarizations is significantly different (Figure 5.4(b)), which indicates that the electronic state of the LSMO conduction channel can actually be adjusted by ferroelectric polarization. To obtain directly the information about magnetic coupling in the heterostructure, magnetic transmission has been carried out through the LSMO layer (Figure 5.4(b)), and the peak position of the magnetoresistance curve is related to the coercive field of the magnetic measurement. Therefore, the mandatory and exchange deviation fields can be derived from these transmission measures. For the upward bias state, the coercivity and exchange bias field are measured at around

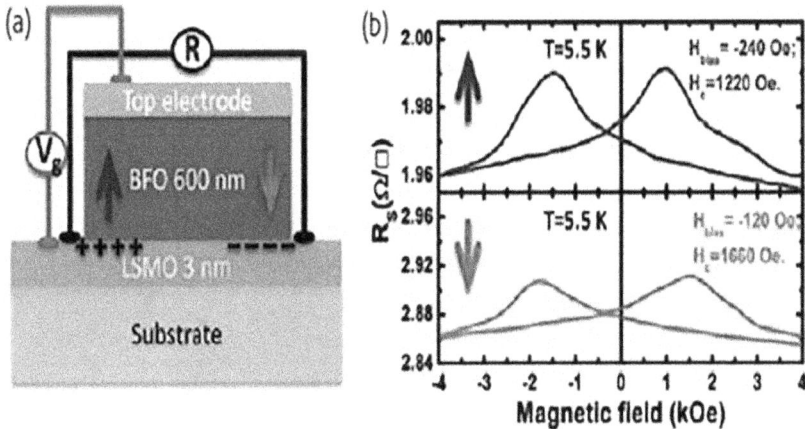

FIGURE 5.4 ME coupling through BFO-LSMO interface, (a) graphic representation of FE FET made up of BFO and LSMO layers, where the arrows depict the out-of-plane FE polarization directions, (b) magneto-transport measurements conceded out with upward (top panel) and upward (bottom panel) FE polarization afterward the gate voltage switching.

Source: Reproduced with permission from Yu et al. (2012)

1220 Oe and 240 Oe, respectively. On the other side, when there is increase in forcing filed (1660 Oe) and the suppression of the exchange bias field (120 Oe), a very different magnetic anisotropy is found for the downward situation. Other measurements also indicate that this effect is reversible. Furthermore, it is controlled by isothermal switching of the ferroelectric polarization of the BFO layer (Chakhalian et al., 2007). These results have proven an important step towards magnetic field control, and may allow the emergence of a new class of controllable spintronic devices, such as multiferroic tunnel junctions, spin valves and tunable spin-polarized two-dimensional electrons. Although the transition temperature of magnetoelectric coupling in the current system is only 100K, through engineering/design of interface magnetic coupling force and relative orbital degrees of freedom, higher temperatures can be achieved in similar systems. From an application point of view, if the transition temperature can be raised above room temperature, it will be a huge improvement.

5.4 CONCLUDING REMARKS

The multifunctional nature of the lead-free magnetic-ferroelectric green composites has a great demand in the current multiferroic materials research due to their potential use in various device applications. In summary of the present investigations, we have addressed the basic understanding of the magnetic-ferroelectric composite formation as per the literature via their structural investigation and also briefly discussed the multiferroic properties i.e., ferroelectric as well as magnetic hysteresis loops. For further confirmation of the multiferroic as well as magnetoelectric nature qualitatively in such composites, P vs E loop measurements under the applied magnetic field are also presented. Magnetoelectric couplings and their use in the various device applications are also discussed using the latest literature available. In conclusion, our investigations provide the basic understanding of the magnetic-ferroelectric composites and their multiferroic properties with the latest applications.

5.4.1 CREDIT AUTHORSHIP CONTRIBUTION STATEMENT

Manish Kumar: Writing—original draft, formal analysis; **Subhash Sharma:** Writing—original draft, formal analysis; **Arvind Kumar:** Writing—original draft, formal analysis; **Abhay Nanda Srivastva:** Formal analysis.

REFERENCES

Bangruwa, J. S., Vashisth, B. K., Singh, N., & Verma, V. (2018). Anomalous ferroelectric and magnetic behavior in BPFO-NZFO multiferroic nano-composites. *Ceramics International*, 44(10), 11737–11744.

Béa, H., Bibes, M., Barthélémy, A., et al. (2005). Influence of parasitic phases on the properties of BiFeO$_3$ epitaxial thin films. *Applied Physics Letters*, 87(7), 072508.

Béa, H., Bibes, M., Fusil, S., Bouzehouane, K., Jacquet, E., Rode, K., & Barthélémy, A. (2006). Investigation on the origin of the magnetic moment of BiFeO$_3$ thin films by advanced x-ray characterizations. *Physical Review B*, 74(2), 020101.

Chakhalian, J., Freeland, J. W., Habermeier, H. U., Cristiani, G., Khaliullin, G., Van Veenendaal, M., & Keimer, B. (2007). Orbital reconstruction and covalent bonding at an oxide interface. *Science, 318*(5853), 1114–1117.

Chandra, P. (2019). Multifunctionality goes quantum critical. *Nature Materials, 18*(3), 197–198.

Chauhan, M., Sanghi, S., & Agarwal, A. (2021). Crystal structure and improved dielectric, magnetic, ferroelectric and magneto-electric properties of x $CoFe_2O_{4-(1-x)}$ $BaTiO_3$ multiferroic composites. *Journal of Materials Science: Materials in Electronics*, 1–18.

Dabas, S., Chaudhary, P., Kumar, M., Shankar, S., & Thakur, O. P. (2019). Structural, microstructural and multiferroic properties of $BiFeO_3$—$CoFe_2O_4$ composites. *Journal of Materials Science: Materials in Electronics, 30*(3), 2837–2846.

Ederer, C., & Spaldin, N. A. (2005). Weak ferromagnetism and magnetoelectric coupling in bismuth ferrite. *Physical Review B, 71*(6), 060401.

Fiebig, M., Lottermoser, T., Meier, D., & Trassin, M. (2016). The evolution of multiferroics. *Nature Reviews Materials, 1*(8), 1–14.

Ghidini, M., Maccherozzi, F., Moya, X., et al. (2015). Perpendicular local magnetization under voltage control in Ni films on ferroelectric $BaTiO_3$ substrates. *Advanced Materials, 27*(8), 1460–1465.

Jain, P., Wang, Q., Roldan, M., Glavic, A., Lauter, V., Urban, C., & Fitzsimmons, M. R. (2015). Synthetic magnetoelectric coupling in a nanocomposite multiferroic. *Scientific Reports, 5*(1), 1–4.

Kumar, M., Shankar, S., Dwivedi, G. D., Anshul, A., Thakur, O. P., & Ghosh, A. K. (2015). Magneto-dielectric coupling and transport properties of the ferromagnetic-$BaTiO_3$ composites. *Applied Physics Letters, 106*(7), 072903.

Lee, O. J., Misra, S., Wang, H., & MacManus-Driscoll, J. L. (2021). Ferroelectric/multiferroic self-assembled vertically aligned nanocomposites: Current and future status. *APL Materials, 9*(3), 030904.

Lee, S., Ratcliff, W., Cheong, S. W., & Kiryukhin, V. (2008). Electric field control of the magnetic state in $BiFeO_3$ single crystals. *Applied Physics Letters, 92*(19), 192906.

Li, Y., Wang, Z., Yao, J., Yang, T., Wang, Z., Hu, J. M., & Viehland, D. (2015). Magnetoelectric quasi-(0-3) nanocomposite hetero structures. *Nature Communications, 6*(1), 1–7.

Lu, X., Kim, Y., Goetze, S., Li, X., Dong, S., Werner, P., & Hesse, D. (2011). Magnetoelectric coupling in ordered arrays of multilayered heteroepitaxial $BaTiO_3$/$CoFe_2O_4$ nanodots. *Nano Letters, 11*(8), 3202–3206.

Matsukura, F., Tokura, Y., & Ohno, H. (2015). Control of magnetism by electric fields. *Nature Nanotechnology, 10*(3), 209–220.

Mo, Z., Tian, G., Yang, W., Ning, S., Ross, C. A., Gao, X., & Liu, J. (2021). Magnetoelectric coupling in self-assembled $BiFeO_3$—$CoFe_2O_4$ nanocomposites on (110)-$LaAlO_3$ substrates. *APL Materials, 9*(4), 041109.

Okabayashi, J., Miura, Y., & Taniyama, T. (2019). Strain-induced reversible manipulation of orbital magnetic moments in Ni/Cu multilayers on ferroelectric $BaTiO_3$. *NPJ Quantum Materials, 4*(1), 1–8.

Ramesh, R., & Spaldin, N. A. (2007). Multiferroics: progress and prospects in thin films. *Nature Materials, 6*(1), 21–29.

Shankar, S., Kumar, M., Ghosh, A. K., Thakur, O. P., & Jayasimhadri, M. (2019). Anomalous ferroelectricity and strong magnetoelectric coupling in $CoFe_2O_4$-ferroelectric composites. *Journal of Alloys and Compounds, 779*, 918–925.

Shankar, S., Kumar, M., Tuli, V., Thakur, O. P., & Jayasimhadri, M. (2018). Energy storage and magnetoelectric coupling in ferroelectric—Ferrite composites. *Journal of Materials Science: Materials in Electronics, 29*(21), 18352–18357.

Sharma, S., Singh, V., Kotnala, R. K., Ranjan, R., & Dwivedi, R. K. (2014). Co-existence of tetragonal and monoclinic phases and multiferroic properties for x $\leqslant$ 0.30 in the (1–x)

Pb(Zr0.52Ti0.48)O3—(x)BiFeO$_3$ system. *Journal of Alloys and Compounds*, *614*, 165–172.

Spaldin, N. A., & Ramesh, R. (2019). Advances in magnetoelectric multiferroics. *Nature Materials*, *18*(3), 203–212.

Tian, G., Zhang, F., Yao, J., Fan, H., Li, P., Li, Z., & Liu, J. (2016). Magnetoelectric coupling in well-ordered epitaxial BiFeO$_3$/CoFe$_2$O$_4$/SrRuO$_3$ hetero structured nanodot array. *ACS Nano*, *10*(1), 1025–1032.

Tokura, Y., Kawasaki, M., & Nagaosa, N. (2017). Emergent functions of quantum materials. *Nature Physics*, *13*(11), 1056–1068.

Wang, Y., Hu, J., & Nan, C. (2010). Multiferroic magnetoelectric composite nanostructures. *NPG Asia Materials, 2*(2), 61–68.

Yu, P., Chu, Y. H., & Ramesh, R. (2012). Oxide interfaces: Pathways to novel phenomena. *Materials Today*, *15*(7–8), 320–327.

6 Hybridization of Reinforced Fibers on the Performance of Green and Eco-Friendly Composites

Brijesh Gangil, Shashikant Verma, Lalit Ranakoti, and Sandeep Kumar

CONTENTS

6.1   INTRODUCTION

In order to create hybrid composite materials, multiple types of fibers are mixed together in a common matrix. Composites made with more than one type of reinforcement offer a broader variety of qualities than traditional composites. At its core, a composite material is one that is made up of at least two or many separate materials. Growth in composite applications can be attributed to a variety of factors, but a key driver is the increased strength and light weight of goods made from composites (Potluri, R. 2019), (Mochane et al., 2019). Composite materials are now used in almost every industry, making it nearly impossible to identify one that doesn't. During the past three to four decades, technological advancements have been significant. As a result of the development of new materials and industrial techniques, new requirements and opportunities have developed with the changing environment. An increasing number of new manufacturing methods and materials

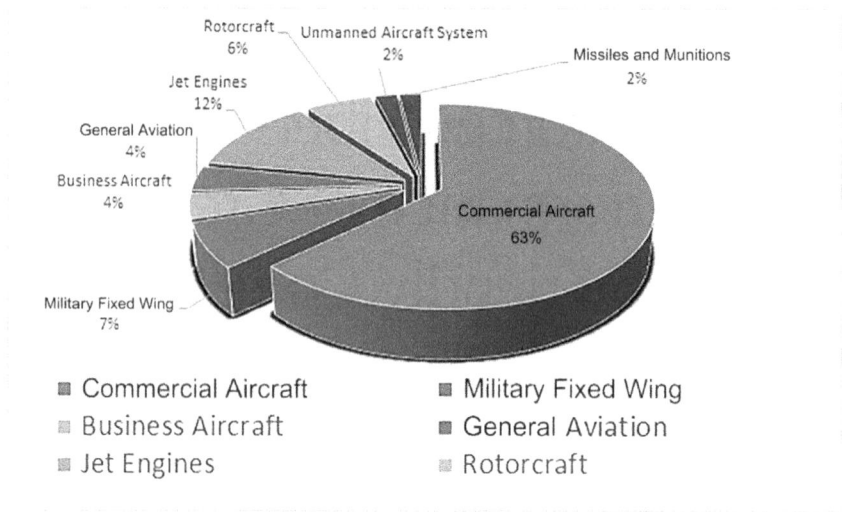

FIGURE 6.1 Estimated 2013–2022 market for aerospace composite structure.

systems have been created during the past decade to meet the demands of various market segments. All of these industries are now heavily relying on the composite materials as shown in **Figure 6.1**. This trend of increasing composites use may be witnessed in particular in the automobile industry, where manufacturers are seeking to improve vehicle economy by reducing the weight of the components that are used. Because of the multiple advantages it provides, the field of polymer-based composite materials draws the attention of a huge number of scientists and researchers from all across the globe (Mosiewicki & Aranguren, 2013), (Khalid et al., 2021).

These characteristics are comprised of the highest strength to weight ratio possible, tribological capabilities, high stiffness to weight ratios, the degree of customizability accessible for a particular application, chemical/corrosion resistance, and the ease with which they may be manufactured.

One of the most pressing issues facing today's generation is environmental preservation. It is more important than ever to develop environmentally friendly products or methods of protecting the environment. Fibers made from natural materials have gained a lot of attention as an alternative to synthetic fibers in advanced applications because they have qualities such as low density and high specific strength (Rajak et al., 2019). Natural fibers provide a number of advantages over synthetic fibers, including low cost, abundance, and the ability to absorb carbon dioxide, which reduces pollution. Natural fibers do not produce toxic fumes during processing, and they are not abrasive to processing machinery. Natural fibers are hampered in polymer reinforcement by their inherent hydrophilic nature and high flammability. High water uptake capacity, poor adhesion at the fiber-matrix interface, and poor fiber dispersion are all outcomes of their hydrophilic nature. Natural fiber chemical modification has been employed as the subject of extensive research in an effort to

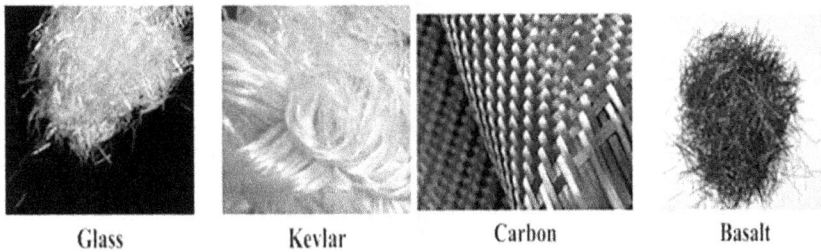

Glass   Kevlar   Carbon   Basalt

FIGURE 6.2   Contribution of countries in hybridized composites.

circumvent these drawbacks (Das, 2017). The overall qualities of fabricated composite materials have been successfully improved by the chemical modification of fibers. However, in another study it was recently reviewed that flame resistance of fiber can be improved by making use of flame retardant technique (Bhagabati, 2020; Gurunathan et al., 2015).

There has been a growing interest in integrating two or more fillers/fibers into a single matrix to develop hybrid composites. Since 2013, there has been a considerable increase in the number of papers pertaining to hybrid composites, as shown in **Figure 6.2**. The primary goal was to overcome the constraints of a single fiber/filler reinforced matrix by incorporating different fibers/fillers with properties that were comparable to or better than the initial fiber/filler. It is a new technology that allows more than one filler/fiber to be mixed into a similar resin in order to overcome a lack of other fillers/fibers in a given area. Some countries are more active in using hybrid composites than others (Faruk et al., 2014; Saba et al., 2014).

6.2   SYNTHETIC FIBERS

Man-made fibers are also known as synthetic fibers. They are synthesized from polymers that aren't readily available in nature. They are primarily derived from fossil fuels or from petrochemicals and are invented in laboratories and at a manufacturing scale. In the comparison of natural fibers, synthetic fibers are deserving of particular interest because of their better properties. Natural fiber quality depends on many factors, including growth conditions, harvesting methods, and maturity. There are a number of disadvantages to synthetic fibers, including their accumulation in the environment and landfills, their price, and the fact that they come from fossil fuels. Therefore, hazardousness of these fibers on the environment is very high, and, for traditional synthetics, the image is far from natural fibers (Prashanth et al., 2017). Synthetic fibers are basically derived from petrochemical, and their production depends on gas reserves and the decline of oil results in them being non-renewable. Afterward, these fibers are non-biodegradable, production based on energy-intensive, and difficult to recycle. Some different types of synthetic fibers are Kevlar fiber, glass fiber, and carbon fiber, as shown in **Figure 6.3**.

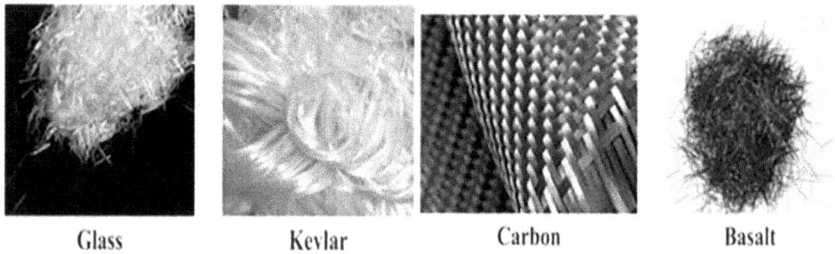

| Glass | Kevlar | Carbon | Basalt |

FIGURE 6.3 Types of synthetic fiber.

6.3 KEVLAR FIBER

These fibers are also referred to as aramid fibers. This compound is generated when para-phenylenediamine and terephthaloyl chloride are mixed, forming aromatic polyamide threads. The fiber strands are known as Kevlar. These fibers are used to make body armor.

6.4 GLASS FIBERS

These fibers are made by drawing liquified glass with 50% silica and other minerals oxides through a procedure. In addition to glass fiber, polymers boron and carbon fibers have mechanical properties that are similar to those of glass fiber. Although not as stiff and strong as carbon fiber, it is far less expensive and brittle when used in composites. As a consequence of this development, glass fibers are used as a reinforcing agent in a variety of polymer products to produce glass-reinforced plastic (GRP), also known as "fiberglass," which is an extremely strong and comparatively lightweight fiber-reinforced polymer (FRP) composite material. GRP, in contrast to glass wool, contains little or no air or gas, is denser, and hence performs poorly as a thermal insulator; instead, it is used structurally because of its strength and low weight; nonetheless, it is more expensive than glass wool. Glass fibers are the most common type of synthetic/artificial fiber (Sathishkumar et al., 2014). They're commonly found in bulletproof glasses, gasoline tanks, vehicle bodywork, and the energy sector, among other places.

6.5 CARBON FIBERS

They range in size from 5 to 10 micrometers in diameter and are mostly made of carbon atoms. Carbon fibers are created from a variety of materials, with polyacrylonitrile (PAN) accounting for about 90% of total production and rayon of petroleum pitch accounting for the remaining 10%. Organopolymers such as PAN and rayon are organic polymers made up of long chains of molecules that are bonded together by carbon atoms. Carbon fibers are routinely combined with other materials in order to create composites (Liu et al., 2018). Carbon fiber reinforced composites are a suitable material for a variety of applications requiring demanding mechanical and weight characteristics due to their qualities. Carbon fibers, because of their light weight, have a high strength-to-weight ratio and a high level of strength and durability. Carbon fibers are also conductive and corrosion-resistant, with a high E-modulus and minimal thermal expansion. Carbon fiber also has a lower density than steel, which makes it perfect for lightweight applications.

6.6 BASALT FIBER

This material exhibits superior physical and mechanical qualities when compared to glass fiber. It is cheaper than carbon fiber. Basalt fiber is produced by melting derivatives of basalt rocks obtained from quarries. These fibers are obtained with a diameter between 10 and 20 µm when extrusion takes place via a tiny nozzle at a temperature of 1500 degrees Celsius (Cheng et al., 2018). Their specific strength is three times greater than that of steel. When it comes to textiles, thin fibers are typically employed to make woven fabrics. Filament winding uses thicker fiber, such as in the manufacture of CNG cylinders or pipes, etc. When it comes to making pultrusion, geogrid, unidirectional fabric, and multiaxial fabric, the most viscous fibers are used (Liang & Hota, 2013). Using continuous basalt fiber to make basalt rebar is one of the best ways to use it. It's also the most popular thing to do in the building industry right now (Inman et al., 2017).

6.7 NATURAL FIBER

Plant or animal-derived biopolymers, such as natural fibers, are used to make biopolymers. Natural fibers derived from plants are mostly composed of cellulose, while natural fibers derived from animals are primarily composed of proteins. Because of its biodegradability, cheap cost, relatively high strength, low density, good acoustical damping, low manufacturing energy consumption, and minimal carbon impact on the environment, plant-based natural fiber is a preferred material in the automobile industry (Ranakoti et al., 2021; Verma et al., 2019a). Some of the few natural fibers that are mostly used are shown in **Figure 6.4**. They are effortless to obtain and abundant in nature. Natural fiber

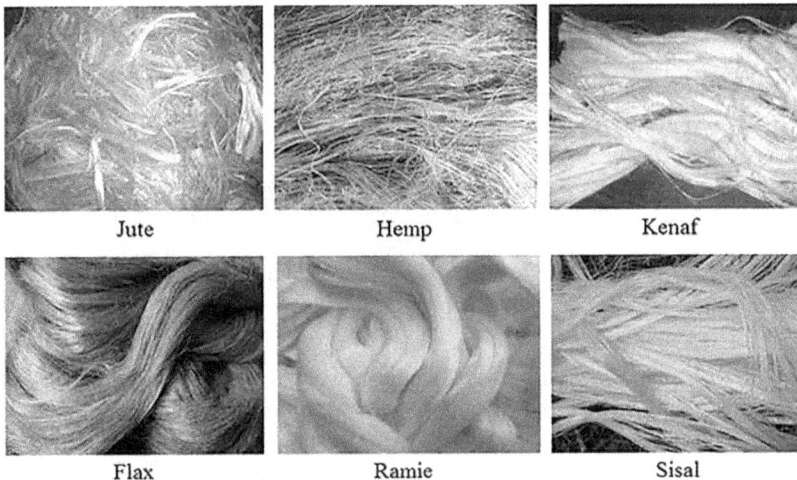

Jute   Hemp   Kenaf

Flax   Ramie   Sisal

FIGURE 6.4   Types of natural fibers.

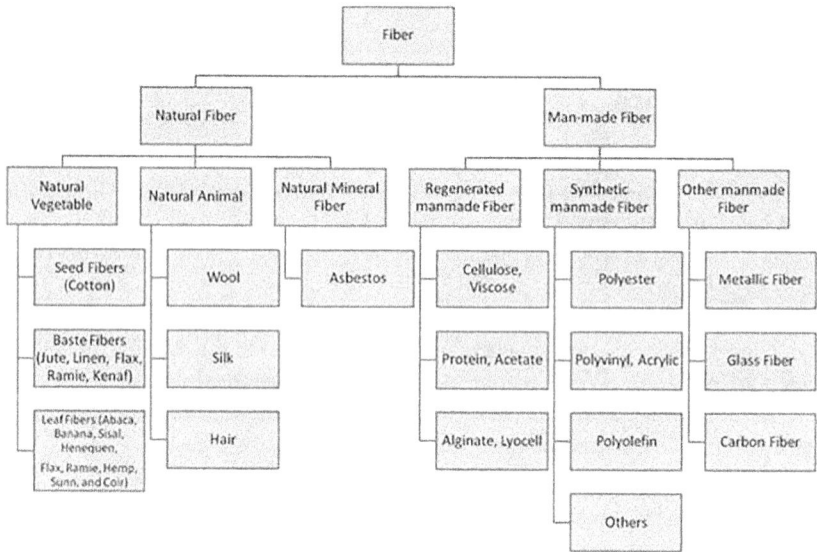

FIGURE 6.5 Classifications of fiber.

composites are preferred over synthetic fiber-reinforced composites for modern applications due to their cost and environmental advantages. Natural fibers' high strength and mechanical properties have motivated materials scientists to use a significant number of natural fibers as reinforcements into composite materials, which have improved their performance. **Figure 6.5** depicts the classification of natural and synthetic fibers. Overall, natural fibers have lower stiffness and strength when compared to synthetic fibers (Gangil et al., 2020). With time, the fiber's properties deteriorate. Fibers that are younger tend to be stronger and more elastic than the older ones, because of their viscoelastic nature, many natural fibers are strain-rate sensitive (Pandey et al., 2010; Kerni et al., 2020). Moreover, natural fiber properties are also influenced by the moisture content of the fiber.

6.8   HYBRIDIZATION

Hybridization is the process of combining two or more kinds of materials into a single resin in order to generate new material in the form of composite laminas and stacked plies that are made up of multiple types of composite materials, as well as new applications. In order to achieve the Hybridization of fiber-reinforced polymer composites, one reinforcing phase must be present.

- A hybrid composite is made with a single matrix and two or more different types of reinforcing fibers or filler agents inside (two or more reinforcing phases).
- The Hybridization can be obtained by consisting of one type of reinforcing agent (fiber/filler) and two or more matrix materials.
- Multiple matrix phases with more than one reinforcing phase are included in the model.

Hybridization when done on natural fiber-based composite, Hybridization when done on natural fiber-based composite, several times the Hybridization of natural-glass fibers have been investigated, whereas polymer matrix material, which may be either synthetic or natural in origin (Sanjay et al., 2015). For advanced application, Hybridization plays an important role and provides a new opportunity to extend the applicability of the composite materials.

6.9   SYNTHETIC-NATURAL FIBERS REINFORCED HYBRID COMPOSITES

The market for composite applications is shifting because of the advent of novel biodegradable polymers. Fiber-based composite having reinforcing (natural) agents such as kenaf, banana, flax, hemp, bamboo, sisal, and jute are becoming more popular and used in Hybridization with synthetic fiber in packaging, sports, automotive, and other industrial sectors. According to (Kumar et al., 2017), synthetic-natural fibers are combined together to make hybrid composites and are used to enhance the performance of resulting material, such as increasing the water uptake resistance, reducing the negative environmental impact, and balancing the fiber's costs. Glass fiber is one of the widely used synthetic fibers in various sectors due to their light weight, strength, and resilient behavior. The combined effect of glass and OPEFB (oil palm empty fruit bunch) hybrid fiber in Phenolformaldehyde (PF) composites is investigated and it is observed that the fabricated composites are cost-effective, light in weight, and have good performance characteristics (Sreekala et al., 2002). These composites could be used as structural materials in situations where strength and affordability are critical. The tensile characteristics of hybrid flax and glass fiber-based composites depend on the glass fiber content and are enhanced by the increment of glass fiber. Hybrid composites made of flax and glass fiber have better interlaminar shear strength and interlaminar fracture toughness than GFRP (Zhang et al., 2013). The Hybridization of synthetic glass fibers and sisal fibers was studied by (Palanikumar et al., 2016). Sisal/glass Hybridization provides intermediate qualities between glass and sisal, according to the study. As a result of Hybridization, the composite's tensile, flexural, and impact strengths increased, but its modulus of elasticity and tensile strength did not change significantly. Serval authors reported on the combined effect of jute-glass fiber in composites and found that the incorporation of glass fiber into polyester composites reinforced with jute fibers resulted in enhanced tensile and flexural characteristics but also reduced water absorption. Synthetic carbon fibers and sisal fibers were combined in a polymer matrix by (Khanam & AlMaadeed, 2014). The hybrid composites have greatly improved their tensile, flexural, and chemical resistance properties compared to their unhybridized counterparts (Verma et al., 2019b, 2021).

6.10   NATURAL-NATURAL FIBERS REINFORCED HYBRID COMPOSITES

Natural fibers are predicted to play a large part in the expanding "green" economy, which is focused on energy efficiency, the use of renewable resources in polymer goods and industrial processes that minimize carbon emissions, and recyclable materials that reduce waste generation. Since the dawn of time, nature and human innovation have continually replenished our supply of natural fibers.

The different Hybridization of natural-natural fibers, jute-hemp-epoxy and jute-hemp-flax-epoxy, and its effect on tribomechanical and dynamic behavior is investigated by (Chaudhary et al., 2018). They revealed that the positive influence of combined fibers on wear behavior of all fabricated composites and minor effects of normal load and speed on coefficient of friction had been observed. Jute-hemp combination is attaining marginal improvement in tribomechanical property among all fabricated composites. (Kumar et al., 2019) prepared the sisal-bauhinia vahlii fiber epoxy composites. The finding showed that the inferior physical and water absorption results are obtained with an increment of mono and hybrid fiber content in epoxy composites, whereas mechanical and wear properties give superior results for hybrid (sisal-bauhinia vahlii) fiber as compared to mono (sisal/bauhinia vahlii) fiber-based composites. Bajpai et al., 2013) investigated the effect of natural fibers on PLA-based composites, and for this purpose, the investigators used grewia optiva, sisal, and nettel fiber as a reinforcing agent. They observed that the incorporation of fibers in PLA composites is decreasing the wear rate and 70% reduction in wear rate are detected as compared to neat PLA composites. In other Hybridization, the incorporation of hemp fiber and eggshell in epoxy composites and its effect on tribological and mechanical properties are investigated by (Inbakumar & Ramesh, 2018). A simple hand layup technique is used to fabricate the composites specimens having different weight proportions of fiber; hemp/eggshell (30%/0.25%; 40/0.50; 50%/1%). They observed that the prepared composite with hemp/eggshell (50%/0.50%) gives superior mechanical properties. The synergistic impact of nettle and bauhinia vahlii fiber in epoxy composites was investigated, as well as the mechanical and tribological parameters of the composites were determined. The higher 6 wt.% based composites with 3%/3% (nettle/bauhinia vahlii) is exhibiting superior in tensile strength (34.05MPa), flexural strength (42.45 MPa), and hardness (37.01 H_v), respectively (Kumar et al., 2020). (Gupta et al., 2018) studied the Hybridization of jute and sisal fibers. The storage and loss modulus values of the sisal/jute Hybridization were very high. The hybrid composite's damping properties have decreased compared to the non-hybrid composite. The hybrid composite absorbed less water than the unhybridized composite. (Jawaid et al., 2011) investigated a tri-layer hybrid composite combining OPEFB/woven jute fiber reinforced in epoxy matrix material as well as other materials. When compared to the mono OPEFB fiber-based epoxy composite, the tensile strength of the hybrid fiber-based composite is much higher. The combined effect of coir-banana fiber-based propylene composites is focused by the Singh and Mukhopadhyay (2020) and revealed that the hybrid (coir-banana) and fiber content is attaining a significant role in sound insulation up to a certain limit. When the same fiber loading was used in both the hybrid coir-banana based hybrid composite and the individual fiber based composite, they discovered that the hybrid coir-banana-based hybrid composite achieved a lower transmission loss.

From the preceding, we can conclude that hybrid composites fabricated with natural (green) fibers have exhibited great potential for use in automobiles, food packaging, agriculture, biomedical building, and residential applications (Kumar et al., 2020).

6.11 CONCLUSION

On the performance of green and eco-friendly composites, a lot of studies are being done on the Hybridization of reinforced fibers. Green composites appear to be an eco-friendly and long-lasting alternative to composites made with typical synthetic fibers and matrixes. Hybridization has gained popularity due to the better performance of the produced products and the capacity to overcome the limits that prevent natural fibers from being used in industrial production systems. The option of incorporating a significant proportion of natural fibers into a standard synthetic reinforced composite product exists, and this would represent a significant step forward from both an ecological and economic standpoint. Research in the domain of Hybridization of reinforced fiber is being conducted in a variety of other sectors as well, with the goal of generating unique pure and hybrid composites that may be used in a variety of applications.

REFERENCES

Bajpai, P. K., Singh, I., & Madaan, J. (2013). Tribological behavior of natural fiber reinforced PLA composites. *Wear*, *297*(1–2), 829–840.

Bhagabati, P. (2020). Biopolymers and biocomposites-mediated sustainable high-performance materials for automobile applications. In *Sustainable Nanocellulose and Nanohydrogels from Natural Sources* (pp. 197–216). London: Elsevier.

Chaudhary, V., Bajpai, P. K., & Maheshwari, S. (2018). Studies on mechanical and morphological characterization of developed jute/hemp/flax reinforced hybrid composites for structural applications. *Journal of Natural Fibers*, *15*(1), 80–97.

Cheng, Y., Wang, W., Gong, Y., Wang, S., Yang, S., & Sun, X. (2018). Comparative study on the damage characteristics of asphalt mixtures reinforced with an eco-friendly basalt fiber under freeze-thaw cycles. *Materials*, *11*(12), 2488.

Das, S. (2017). Mechanical properties of waste paper/jute fabric reinforced polyester resin matrix hybrid composites. *Carbohydrate Polymers*, *172*, 60–67.

Faruk, O., Bledzki, A. K., Fink, H. P., & Sain, M. (2014). Progress report on natural fiber reinforced composites. *Macromolecular Materials and Engineering*, *299*(1), 9–26.

Gangil, B., Ranakoti, L., Verma, S., Singh, T., & Kumar, S. (2020). Natural and synthetic fibers for hybrid composites. In *Hybrid Fiber Composites: Materials, Manufacturing, Process Engineering* (pp. 1–15). New York: Wiley.

Gupta, M. K., Choudhary, N., & Agrawal, V. (2018). Static and dynamic mechanical analysis of hybrid composite reinforced with jute and sisal fibres. *Journal of the Chinese Advanced Materials Society*, *6*(4), 666–678.

Gurunathan, T., Mohanty, S., & Nayak, S. K. (2015). A review of the recent developments in biocomposites based on natural fibres and their application perspectives. *Composites Part A: Applied Science and Manufacturing*, *77*, 1–25.

Inbakumar, J. P., & Ramesh, S. (2018). Mechanical, wear and thermal behaviour of hemp fibre/egg shell particle reinforced epoxy resin bio composite. *Transactions of the Canadian Society for Mechanical Engineering*, *42*(3), 280–285.

Inman, M., Thorhallsson, E. R., & Azrague, K. (2017). A mechanical and environmental assessment and comparison of basalt fibre reinforced polymer (BFRP) rebar and steel rebar in concrete beams. *Energy Procedia*, *111*, 31–40.

Jawaid, M., Khalil, H. A., & Bakar, A. A. (2011). Woven hybrid composites: Tensile and flexural properties of oil palm-woven jute fibres based epoxy composites. *Materials Science and Engineering: A*, *528*(15), 5190–5195.

Kerni, L., Singh, S., Patnaik, A., & Kumar, N. (2020). A review on natural fiber reinforced composites. *Materials Today: Proceedings*, *28*, 1616–1621.

Khalid, M. Y., Al Rashid, A., Arif, Z. U., Ahmed, W., Arshad, H., & Zaidi, A. A. (2021). Natural fiber reinforced composites: Sustainable materials for emerging applications. *Results in Engineering*, *11*, 100263.

Khanam, P. N., & AlMaadeed, M. A. (2014). Improvement of ternary recycled polymer blend reinforced with date palm fibre. *Materials & Design*, *60*, 532–539.

Kumar, S., Gangil, B., Mer, K. K. S., Gupta, M. K., & Patel, V. K. (2020). Bast fiber-based polymer composites. *Hybrid Fiber Composites: Materials, Manufacturing, Process Engineering*, 147–167.

Kumar, S., Mer, K. K. S., Gangil, B., & Patel, V. K. (2019). Synergy of rice-husk filler on physico-mechanical and tribological properties of hybrid Bauhinia-vahlii/sisal fiber reinforced epoxy composites. *Journal of Materials Research and Technology*, *8*(2), 2070–2082.

Kumar, S., Mer, K. K. S., Gangil, B., & Patel, V. K. (2020). Synergistic effect of hybrid Himalayan Nettle/Bauhinia-vahlii fibers on physico-mechanical and sliding wear properties of epoxy composites. *Defence Technology*, *16*(4), 762–776.

Kumar, S., Prasad, L., & Patel, V. K. (2017). Effect of Hybridization of glass/kevlar fiber on mechanical properties of bast reinforced polymer composites: A review. *American Journal of Polymer Science & Engineering*, *5*(1), 13–23.

Liang, R., & Hota, G. (2013). Fiber-reinforced polymer (FRP) composites in environmental engineering applications. In *Developments in Fiber-reinforced Polymer (FRP) Composites for Civil Engineering* (pp. 410–468). London: Woodhead Publishing.

Liu, H., Dai, Z., Cao, Q., Shi, X., Wang, X., Li, H., & Zhou, J. (2018). Lignin/polyacrylonitrile carbon fibers: The effect of fractionation and purification on properties of derived carbon fibers. *ACS Sustainable Chemistry & Engineering*, *6*(7), 8554–8562.

Mochane, M. J., Mokhena, T. C., Mokhothu, T. H., Mtibe, A., Sadiku, E. R., Ray, S. S., & Daramola, O. O. (2019). Recent progress on natural fiber hybrid composites for advanced applications: A review. *Express Polymer Letters*, *13*(2), 159–198.

Mosiewicki, M. A., & Aranguren, M. I. (2013). A short review on novel biocomposites based on plant oil precursors. *European Polymer Journal*, *49*(6), 1243–1256.

Palanikumar, K., Ramesh, M., & Hemachandra Reddy, K. (2016). Experimental investigation on the mechanical properties of green hybrid sisal and glass fiber reinforced polymer composites. *Journal of Natural Fibers*, *13*(3), 321–331.

Pandey, J. K., Ahn, S. H., Lee, C. S., Mohanty, A. K., & Misra, M. (2010). Recent advances in the application of natural fiber-based composites. *Macromolecular Materials and Engineering*, *295*(11), 975–989.

Potluri, R. (2019). Natural fiber-based hybrid bio-composites: Processing, characterization, and applications. In *Green Composites* (pp. 1–46). Singapore: Springer.

Prashanth, S., Subbaya, K. M., Nithin, K., & Sachhidananda, S. (2017). Fiber reinforced composites-a review. *Journal of Material Sciences & Engineering*, *6*(3), 2–6.

Rajak, D. K., Pagar, D. D., Menezes, P. L., & Linul, E. (2019). Fiber-reinforced polymer composites: Manufacturing, properties, and applications. *Polymers*, *11*(10), 1667.

Ranakoti, L., Rakesh, P. K., & Gangil, B. (2021). Effect of tasar silk waste on the mechanical properties of jute/grewia optiva fibers reinforced epoxy laminates. *Journal of Natural Fibers*, 1–13.

Saba, N., Tahir, P. M., & Jawaid, M. (2014). A review on potentiality of nano filler/natural fiber filled polymer hybrid composites. *Polymers*, *6*(8), 2247–2273.

Sanjay, M. R., Arpitha, G. R., & Yogesha, B. (2015). Study on mechanical properties of natural-glass fibre reinforced polymer hybrid composites: A review. *Materials Today: Proceedings*, *2*(4–5), 2959–2967.

Sathishkumar, T. P., Satheeshkumar, S., & Naveen, J. (2014). Glass fiber-reinforced polymer composites—A review. *Journal of Reinforced Plastics and Composites*, *33*(13), 1258–1275.

Singh, V. K., & Mukhopadhyay, S. (2020). Studies on the effect of Hybridization on sound insulation of coir-banana-polypropylene hybrid biocomposites. *Journal of Natural Fibers*, 1–10.

Sreekala, M. S., George, J., Kumaran, M. G., & Thomas, S. (2002). The mechanical performance of hybrid phenol-formaldehyde-based composites reinforced with glass and oil palm fibres. *Composites Science and Technology*, *62*(3), 339–353.

Verma, S. K., Gangil, B., Gupta, A., Rajput, N. S., & Singh, T. (2021). Dolomite dust filled glass fiber reinforced epoxy composite: Influence of fabrication techniques on physicomechanical and erosion wear properties. *Polymer Composites*, 1–15.

Verma, S. K., Gupta, A., Patel, V. K., Gangil, B., & Ranikoti, L. (2019a). The potential of natural fibers for automotive sector. In *Automotive Tribology* (pp. 31–49). Singapore: Springer.

Verma, S. K., Gupta, A., Singh, T., Gangil, B., Jánosi, E., & Fekete, G. (2019b). Influence of dolomite on mechanical, physical and erosive wear properties of natural-synthetic fiber reinforced epoxy composites. *Materials Research Express*, *6*, 125704.

Zhang, Y., Li, Y., Ma, H., & Yu, T. (2013). Tensile and interfacial properties of unidirectional flax/glass fiber reinforced hybrid composites. *Composites Science and Technology*, *88*, 172–177.

7 Fatigue Phenomenon in Natural Fiber Composites

*Himanshu Prajapati, Anurag Dixit,
and Abhishek Tevatia*

CONTENTS

7.1   INTRODUCTION

The advanced technologies affect manufacturing economics and market capitalization. These technologies are an integral part of today's applications. One of the best examples of advanced technology in the field of material science is the development of natural fiber composites (NFCs) which are the best environmentally friendly material that can help to improve the environmental quality and the economics of the market (Al-Oqla & Salit, 2017). In various industrial applications, the NFCs can be used as a substitute for various non-biodegradable materials (Pickering, 2008). On contract, the glass fibers reinforced composites (GFRC) have non-biodegradable composites that make them hard to even recycle. GFRC eventually generates harmful gases, thus driving our ecosystem to a noxious environment (Case & Reifsnider, 2003; Priyanka et al., 2017). Natural fibers are abundant on earth, and most, after prolonged biodegradation, end up in the soil as humus and release non-toxic gases. Most natural fibers are derived from plants and, with adequate strength and durability, can be used as the main reinforcement material for the production of NFCs (Zimniewska et al., 2011).

Repeated loading and unloading of a material is common in daily life, for example a vehicle driving on a road with several bumps, a public chair, opening and closing doors, shoes and knife handles (Sangrey et al., 1969; Wasage et al., 2010). The studies of the fatigue of metallic structures have been carried out for more than 190 years. Therefore, researchers thoroughly understand the mechanism of fatigue in metals (Forrest, 2013). The fatigue damage occurs cumulatively, causing sudden ruptures and failure (Talreja & Varna, 2015). Nonetheless, evaluating the fatigue life of components at an early stage is very important so that maintenance or replacement can be planned

DOI: 10.1201/9781003272625-7

83

well in advance against a catastrophic failure. The fatigue life analysis must include the prediction of the remaining life of these structures (Nishida, 2014).

After the 1980s, researchers started comparing the fatigue behavior of composites to metals (Miraftab, 2009). The composite structures represent the sound fatigue-resistant properties during the service period (Muthukumar, 2019). The continuous technological advancements explored the major factors influencing the fatigue life of composite, such as level of stress inside the material, area of the reinforcements, the incompatibility stiffness and matrix locking stresses during the manufacturing/loading (El Messiry, 2017; Miraftab, 2009). The nucleation or initiation of the crack inside the composite material promotes damage evolution which leads to the short crack formation and successively long crack and at last leading to the catastrophic failure of material (Bruzzi & McHugh, 2004; Li & Ellyin, 2000; McDowell, 1997).

The concept of fracture mechanics is widely applied to evaluate the critical size of a crack (Beevers, 1977; Broek, 2012; Carlson et al., 2012). The micro-structural features such as constraints of fiber in the matrix, reinforcement volume fraction, interfacial bonding strength, particle shape and size play an important role in the evolution of fatigue damage in composite materials (Ding et al., 2002, 2003; Tevatia & Srivastava, 2018; Tevatia & Srivastava, 2015). It has also been observed that the knowledge of strained controlled fatigue damage tolerance characteristics is required for deeply studying the fatigue behavior of composites (Wu et al., 2016).

The research studies carried out so far on various combinations of natural fiber composite materials under fatigue loading are shown in the pie chart in Figure 7.1 (Gassan, 2002; Yuanjian & Isaac, 2007; Towo & Ansell, 2008a, 2008b; Abdullah et al., 2012; Katogi et al., 2012; Shah et al., 2013a; Liang et al., 2012, 2014; de Vasconcellos et al., 2014; El Sawi et al., 2014; Fotouh, 2014; Fotouh et al., 2014; Mir et al., 2014; Hojo et al., 2014; Shahzad & Isaac, 2014; Asgarinia et al., 2015; Ueki et al., 2015; Ismail & Aziz, 2015; Shah, 2016; Bensadoun, 2016; Dobah et al., 2016; Bensadoun et al., 2016; Padmaraj et al., 2017). The flax-epoxy composites are the first choice of researchers in fatigue analysis. Flax is the strongest fiber in all the natural fibers. Flax fiber reinforced composites (FFRC) are excellent absorbers of vibrations (Prajapati et al., 2021). A higher load is required in FFRC to achieve the same damage in other natural fibers as in jute reinforcements. FFRC have lower degradation rates than other composites, even GFRC.

This chapter describes a comprehensive overview of scientific and technical concepts on the fatigue behavior of NFCs and the immense research and development potential

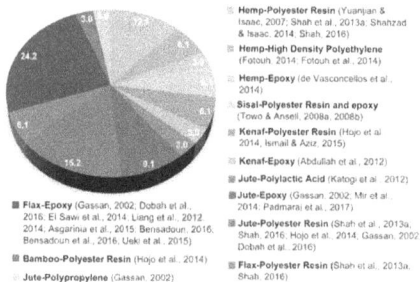

FIGURE 7.1 The number of research work on fatigue analyses of natural fiber composite.

in this domain. The fatigue behavior of NFCs requires an understanding of both the theoretical and experimental aspects of fatigue life estimation. The chapter focuses on the fatigue micro-mechanisms of NFCs under cyclic and variable amplitude loading. Most of the work mentioned in this chapter comprises flax, hemp, sisal, and jute.

7.2 FATIGUE ANALYSIS AND ITS DESIGN

Fatigue or dynamic loads lead to alternative fluctuations in the applied strain and the load on the component. When a certain task is carried out repeatedly, components are subject to different mechanical loads which, under certain conditions, leads to premature failure. Hence, any variable load is called a kind of fatigue load. An alternative stress occurs when stresses or stresses in a component change over time, i.e., the load changes in a certain pattern over time. Various mechanical systems and devices make up the reciprocating or rotating components (Sclater & Chironis, 2001). Fatigue design is a critical factor in designing the mechanical components under dynamic loading. Consequently, a fatigue study is required for the construction of many mechanical or civil structures, e.g. ships, planes, jet fighters, trains, etc. (Hollaway, 2010).

Fatigue fracture occurs in the material at stresses, below the yield point under cyclic loading in a certain frequency range. The fatigue strength indicates the endurance of material at specified alternating loading. The alternative loading is essential for understanding the fatigue failure and inadequacies of the design. There are several analytical and experimental methods available for quantifying this behavior (Osgood, 2013; Roylance, 2001).

In the cases of fatigue potential problems, finite element techniques can also be used to identify stresses (Gokhale, 2008). The simulation of fatigue problems can be performed with appropriate FEA software that requires service loads, dynamic stress-strain properties and stress concentration factors. Utilizing this process, it can be used to approximate the magnitude of local stresses and mean stresses. Consequently, fatigue life becomes a vital parameter for designing an engineering structure for a predefined service life with economical factor of safety (Xue et al., 2020).

Fatigue analysis of any mechanical components determines the remaining life of a material. The amount of repeated stress affects the life of the material (Zhou et al., 2012). The ratio of minimum stress, σ_{max} to maximum stress, σ_{min} is known as the stress ratio, which is represented by R as follows:

$$R = \frac{\sigma_{min}}{\sigma_{max}} \tag{1}$$

The frequency of loading is a very important parameter, which, if increased, will reduce the life of the material. The fatigue strength is the critical stress at which the material fails with repeated loading within a certain stress ratio. There are three designing criteria under fatigue loading, namely the Goodman, Soderberg lines, and the Gerber parabola (Senalp et al., 2007). Depending on the required safety factors, the value of the critical parameter is determined for the designing required service life.

In case of Soderberg line, the linear relationship between yield strength against the mean stress (σ_m) and fatigue strength against the alternating stress (σ_a) is considered as shown in Figure 7.2. We define factor of safety using Soderberg line as:

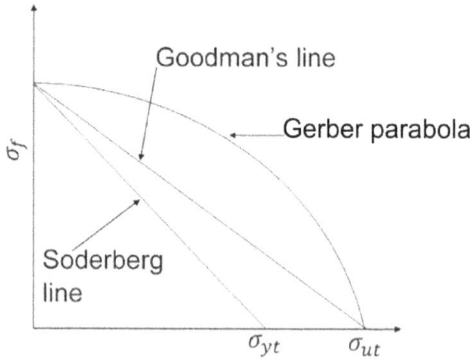

FIGURE 7.2 Soderberg, Goodman, Gerber criterion for fatigue analysis.

$$\frac{1}{fos} = \frac{\sigma_m}{\sigma_{yt}} + \frac{\sigma_a}{\sigma_f} \tag{2}$$

where, *fos* is factor of safety, σ_m is mean stress and equal to $\frac{\sigma_{max} + \sigma_{min}}{2}$, σ_a is alternating stress and equal to $\frac{\sigma_{max} - \sigma_{min}}{2}$, σ_{yt} is yield strength, σ_f is fatigue strength. The Goodman line takes into account the linear relationship between tensile strength (σ_{ut}) versus mean stress and fatigue strength versus alternating stress as:

$$\frac{1}{fos} = \frac{\sigma_m}{\sigma_{ut}} + \frac{\sigma_a}{\sigma_f} \tag{3}$$

The Gerber parabola is a parabolic relationship between ultimate tensile strength (σ_{ut}) against mean stress and fatigue strength against the alternating stress as:

$$\frac{1}{fos} = \left(\frac{\sigma_m}{\sigma_{ut}}\right)^2 + \frac{\sigma_a}{\sigma_f} \tag{4}$$

The analyses are more preferably carried out on cyclic loading. Some studies perform a complete reversal of the loading (Gassan, 2002; Liang et al., 2012, 2014; El Sawi et al., 2014; Asgarinia et al., 2015; Ueki et al., 2015; Dobah et al., 2016; Bensadoun et al., 2016), while others also use monotonous loading where the loading is in one direction but not in the opposite direction of a specific study as applied (Fotouh, 2014; Shah, 2016). However, some efforts have been made to perform a multi-axis fatigue analysis in order to precisely design the required device based on the analysis (Branco et al., 2021).

7.3 FATIGUE DAMAGE EVOLUTION

The fatigue causes damage to the structures when subjected to cyclic loads. Fatigue manifests as microcrack initiation or nucleation, followed by slow crack propagation under cyclic loading and finally failure (Li & Ellyin, 2000; McDowell, 1997). In the early 20th century, Ewing (Ewing & Humfrey, 1903; Orowan, 1939) explained the origin of microcrack under cyclic loading. These microcracks are developed due to harsh environments, material defects, damages in service and discontinuities in design (Barter et al., 2012). The phases of fatigue failure are crack initiation and propagation as shown in Figure 7.3. In metallic structure, the crack initiation period contributed the significant part of total fatigue life and the substantial portion of total fatigue life may be covered during the crack propagation period, under higher stress level (Broek, 1991). On the other hand in composite materials it is vice-versa (Blandford, 2001; McDowell et al., 2003; Newman Jr, 1998). It was very difficult to define the transition from the crack initiation period to the crack propagation period. The concept of fracture mechanics was used to study development of microcrack under fatigue loading (D Broek, 1991; Chell, 1979; Collins, 1993; Dowling, 1993; Ellyin et al., 1997; Laird & Smith, 1963).

Due to manufacturing anomalies or overloading of UD (unidirectional) composite materials, micro-damage to fibers or matrix material can occasionally occur due to axial loading in the direction of the fibers (Zhao et al., 2018). When the UD composite is operated in alternate or dynamic loadings, these microscopic imperfections can lead to initial cracking and propagation in the fiber-matrix interface or in the fiber or matrix itself, which then leads to catastrophic failure of the structure. In the case of long fiber composites with a lower fiber elongation than the matrix elongation, the quasi-static tensile stress in the fiber direction on such composites leads to fiber breakage at random positions along the fiber length, which is due to imperfections and defects in the fibers (Heinecke & Willberg, 2019). This can cause the fiber, like brittle material, to break under an arbitrary load that can be statistically investigated (Yang & Jones, 1980). The Weibull distribution is generally used to statistically investigate this type of failure stress in the case of such a brittle fiber failure for long fiber composites (Curtin & Takeda, 1998; Harlow & Phoenix, 1981; Pitt & Phoenix, 1983).

The increase in tension in UD composite promotes new breaks in the fibers at points where the local tension-strength ratio is higher under cyclic loading condition. Mechanical properties of the fiber and matrix and the strength of the fiber-matrix interface determine the scope for new breaks or cracks in the rest of the fiber/matrix material (Wang et al., 1984). The frame of damage in the fiber/matrix in UD composite is shown in Figure 7.4(a) due to fiber breakage, with the other adjacent fiber, which is stiffer than the broken fiber, remaining under tension, creating a stress concentration in the matrix void created by the fiber break. As a result, the tensions in the matrix are extremely high compared to the fibers, which leads to cracking and spreading in the other directions (Beyerlein & Phoenix, 1997; Laws & Dvorak, 1987).

In Figure 7.4 (b), the matrix is strong enough to prevent the crack from developing on its own. This in turn temporarily stops the strain in the fiber near the crack, but small microcracks continue to form, which damage the matrix, and the propagation of the crack further damages the other fibers. The matrix yields and permanent plastic strains

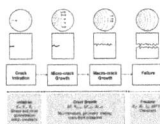

FIGURE 7.3 Mechanism of fatigue crack growth process and its relevant factors.

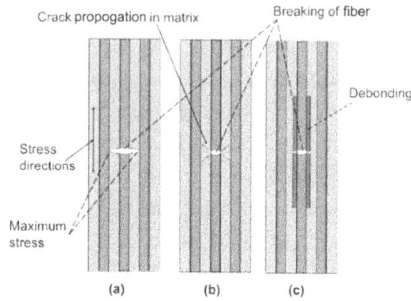

FIGURE 7.4 Framework of crack development in (a) fiber, (b) matrix, (c) fiber matrix interface (debonding) of unidirectional composite.

lead to complete failure of the composite material (Behzadi et al., 2009). In Figure 7.4 (c), the fiber-matrix interface is weak due to incomplete adhesion or chemical reaction between the fiber surface and the matrix, or it may be due to poor chemical treatment. The weak bond between the fiber and matrix interface leads to peeling at various random locations where the bond is weakest (King et al., 1992).

7.4 FATIGUE BEHAVIOR OF NATURAL FIBER COMPOSITE

Like any other composites, natural fiber reinforced composites are also prone to fatigue fractures with changing loads. The composite has completely different fatigue strengths than the starting materials themselves. When NFCs are subjected to fatigue loading after a suitable stress ratio is applied, the NFCs develop microcracks in the matrix material that propagate towards the fibers and a chain of events begins, and finally the sample fails beyond its fatigue limit. The fatigue analysis is carried out considering various natural fibers and matrix materials with unidirectional fiber orientation (Abdullah et al., 2012; Bensadoun et al., 2016; Ezeh & Susmel, 2018, 2019; Katogi et al., 2012; Shah et al., 2013a; Towo & Ansell, 2008b; Ueki et al., 2015). Some of the fatigue studies were carried out with constant stress amplitude, uniaxial tension-tension loading, at a loading frequency of 5 Hz with a loading ratio of 0.1 (Asgarinia et al., 2015; El Sawi et al., 2014; Liang et al., 2012, 2014; Ueki et al., 2015). The parameter values for different NFCs are listed in Table 7.1. Loading frequency can affect the fatigue life of NFCs depending on the orientation of fiber and matrix material (Liang et al., 2012).

The fatigue failure mechanism for natural fiber composites is usually similar to quasi-static stress failure, which can be seen from the microstructural investigation of NFCs (El Sawi et al., 2014; Shah et al., 2013b; Sodoke et al., 2016). Towo & Ansell (2008b) conducted experimental fatigue tests on NaOH-treated sisal-polyester/epoxy composites with R = 0.1 for tensile stress and R = −1 for inverse NaOH stress, while hysteresis loops for R = 0.1 with decreasing area with increasing number of cycles and vice versa for R = −1. Sodoke et al. (2016) evaluated the fatigue characteristics of epoxy composite materials reinforced with flax fibers having $0/90/\pm45^0$ stacking sequences and reported that the composite material had good fatigue strength without aging compared to an aged specimen. Gao et al. (2021) suggested a method utilizing spectroscopic techniques for obtaining the fatigue life of a laminated composite

TABLE 7.1

Fatigue Testing Data for Different NFCs under Fatigue Loading

Fiber	Matrix Material	Orientation	Volume Fraction of Fiber	Tensile Strength (MPa)	Fatigue Strength (MPa)	Porosity	Test Freq.	Load Ratio (R)	Ref.
Flax	Epoxy	0°	43.1±1.5	318	145.9	1–3	5	0.1	(Liang et al., 2012, 2014)
		90°	-	26.1	12.0	-	-	-	(Liang et al., 2012, 2014)
		0°/90°	43.7±1.5	170	78.0	1.3±0.8	5	0.1	(Liang et al., 2012, 2014)
		±45°	42.5±1	79	36.2	3.8±1	5	-	(Liang et al., 2012, 2014)
		0°	48–49	304	139.4	-	5	0.1	(El Sawi et al., 2014)
		±45°	-	68.4	31.4	-	-	-	(El Sawi et al., 2014)
		Twill weave 200g/m²	34.3±0.4	106	48.6	3.7±1.1	5	0.1	(Asgarinia et al., 2015)
		Twill weave 550g/m²	42.3±0.3	105.9	48.6	0.2±0.1	5	-	(Asgarinia et al., 2015)
		Twill weave 224g/m²	31.1±0.7	112.2	51.5	0.2±0.1	1,1.5,3	-	(Asgarinia et al., 2015)
	Polyester	0°	31.4	282	129.4	0–0.6	5, 2	0.1	(Ueki et al., 2015)
		0°	27.7	24.7±26.5	11.3–12.15	0.9±0.3	10	0.1	(Shah et al., 2013)
		±45°	28.9	51.4±2.8	23.6	0.1–0.5	10	0.1	(Shah et al., 2013)
		90°	25.8	12.8–13.6	5.9	0.5–0.9	10	0.1	(Shah et al., 2013)
Hemp	Epoxy	0°/90°	36	113±9	51.4	4	1	0.01	(de Vasconcellos et al., 2014)
		±45°		66±7	30.3	-	-	-	(de Vasconcellos et al., 2014)
	Polyester	Mat	40.84	53	24.3	-	1	0.1	(Yuanjian & Isaac, 2007)
		0°	35.6±0.8	171.3±6.5	78.0	1.3±0.4	10	0.1	(Shah et al., 2013)
		Mat	51.78	46.6±4.6	20.6	-	1	0.1,–1	(Shahzad & Isaac, 2014)
		Mat	14.34	20.1	9.2	-	1	0.1	(Yuanjian & Isaac, 2007)
	HDPE	0°	13.5	29.54±0.18	13.3	-	-	-	(Fotouh et al., 2014)
		0°	30.1	30.18	13.8	-	-	-	(Fotouh et al., 2014)
		0°	3.5		-	-	-	-	(Fotouh et al., 2014)
Sisal	Epoxy	0°	71.5±2.5	329.8±20.9	149.1	-	1.5–3.9	0.1,–1	(Towo & Ansell, 2008a)
		0°	70	329.8±20.9	147.7	-	1.5–3.9	0.1	(Towo & Ansell, 2008b)
	Polyester	0°	68.2±3.2	222.6±21.2	100.9	-	1.5–3.9	0.1,–1	(Towo & Ansell, 2008a)
		0°	70	222.6±21.2	98.95	-	1.5–3.9	0.1	(Towo & Ansell, 2008b)
Jute	Polyester	0°	31.7	175.1±10.3	81.3365	4.2±0.8	10	0.1	(Shah et al., 2013)

material applied with random vibration loads. Some researchers have proposed a simple relationship between the fatigue stress and loading number of cycles which gives satisfactory results in range 10^2 to 10^6 number of cycles as (Mandell et al., 1981):

$$\sigma = \sigma_u - B\left(\log N_f\right) \tag{5}$$

where, B is fatigue strength coefficient, N_f is the cycles of failure.

Figure 7.5 shows the stress vs number of cycle curve various combinations of NFCs available in the literature (Abdullah et al., 2012; Bensadoun et al., 2016; Ezeh & Susmel, 2018, 2019; Katogi et al., 2012; Shah et al., 2013a; Towo & Ansell, 2008b; Ueki et al., 2015). It can be seen that mechanical and microstructural parameters such as fiber volume fraction, fiber type, chemical fiber treatment and matrix material influences the fatigue strength of the composite material. Unidirectional *flax-epoxy* composite having 31.4% volume fraction has the highest fatigue strength among all other volume fractions NFCs as it developed only a few microcracks. Increasing the flax volume fraction to 40% in epoxy resulted in an abrupt decrease in fatigue strength as the microcracks propagated to the matrix material. Therefore, there must be an optimal fraction of the components needed to build a composite material to gain the highest possible fatigue strength. The epoxy polymer has low fatigue strength because it is a brittle material and has lower tensile strength. However, if it is reinforced with some of the glass and natural fibers, its fatigue strength increases.

As shown in Figure 7.5 *kenaf-epoxy* composite having 15% and 45% volume fraction, the fatigue strength will improve by 58.46% and 174.86% as compared with matrix material, respectively. The chemical treatment of fibers also plays a role in improving the mechanical properties of the resulting composite material. The 70% volume fraction of the NaOH-treated *sisal-epoxy* composite exhibits a fatigue strength improvement of 1.82% of the untreated *sisal-epoxy* composite. *Jute-polylactic acid (PLA)* composite with 27% by volume has the highest value for fatigue strength with 48% and 71.89% improvement in fatigue strength compared to the PLA composite reinforced with 28% and 31% by volume, respectively. The PLA has higher strength for low cycle fatigue, but as the number of cycles is further increased, a decrease in fatigue strength is observed, which is less than that of the *jute-PLA* composite for high cycle fatigue.

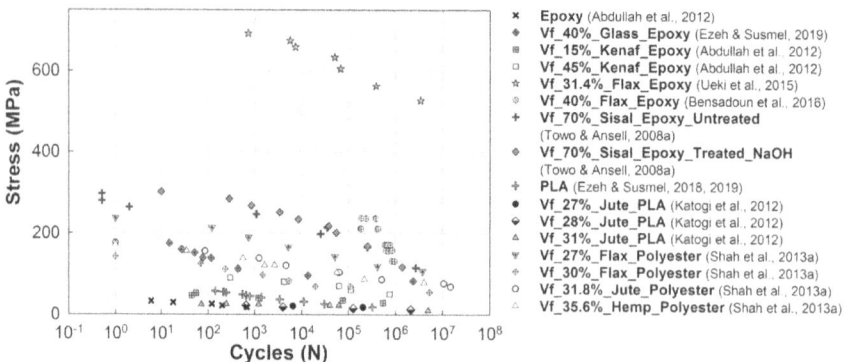

FIGURE 7.5 SN diagram for different unidirectional natural fiber reinforced composite.

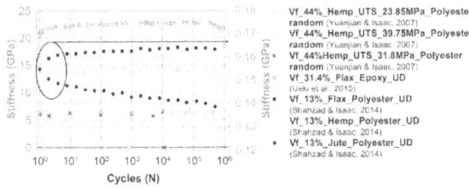

FIGURE 7.6   Stiffness vs cycles for different NFCs.

Flax-polyester composite having 27% volume fraction has the highest fatigue strength of 93.2%, 36.3% and 55.5% compared to *flax-polyester* composite having 30% volume fraction, *hemp-polyester* composite with 35.6% volume fraction and *jute-polyester* composite having 31.8% volume fraction of each, as shown in Figure 7.5. While *hemp-polyester* composite having 31.6% volume fraction has the highest fatigue strength of 14.1% or 41.75% compared to *jute-polyester* composite having 31.8% volume fraction and *flax-polyester* composite having 30% volume fraction, respectively. The comparison of fatigue strength with loading cycles (Figure 7.5) depicted that, except for *flax-epoxy* composites, NFCs fatigue strengths are comparable to glass fiber composites. In fact, *flax-epoxy* composites showed the highest fatigue strengths among other NFCs.

The stiffness gradually increases with the number of cycles for dynamic loads up to a certain cycle range, but decreases abruptly as the number of cycles approaches the corresponding failure cycles, as shown in Figure 7.6 (Shah, 2016; Ueki et al., 2015; Yuanjian & Isaac, 2007). From Figure 7.6, it can be deduced that the flax-reinforced epoxy composite materials have the highest stiffness value among all selected composite materials under the cyclic load. The four composites are shown on the right vertical axis and four are shown on the left side of the vertical axis, as shown in Figure 7.6 as circled and arrowed.

7.5   CONCLUSIONS

Fatigue is a failure of the material below its yield point due to alternating stresses acting on the material. The fatigue analysis provides the life of the material subjected to repeated loads. Fatigue occurs in all components of a composite material and leads to microcracks, delamination and failure.

- The natural fiber composites are more subject to fatigue stress and thus more prone to fatigue failure. The material combination of the composite material influences the fatigue strength of the entire material due to the chemical bond between the natural fibers and the matrix material, clearly visible in the flax-epoxy and flax-polyester composites.
- The stiffness increases to a critical number of cycles. The material is subjected to cyclic loading, the crack propagation is stopped via an obstacle and drops to the lowest value with critical cycles.
- The natural fiber composites are both environmentally friendly and inexpensive. Therefore, the fatigue study will provide various technical data for the development of multi-purpose composites under dynamic loadings.

- The Wöhler curve alone is not enough to determine the overall fatigue properties of NFCs. The SN curve for the NFCs is merely a representation of its resistance to dynamic loading. If a large number of parameters are included in the picture, the SN curve approach proves to be restrictive. The parameters such as varying load amplitude, frequency of the load, type of load, monotonic or cyclical, fiber volume fraction, lamella geometries, thickness, etc. are vital for predicting the fatigue failure of the NFCs. Consideration of such parameters is associated with great uncertainty, which requires new formulations and rules for studying the characteristics of the NFCs acting under cyclic loading. A logical and systematic framework is required for the investigation of fatigue characteristics of NFCs having the provision of the fatigue life diagram, which also takes into account the other parameters as well.

REFERENCES

Abdullah, A. H., Alias, S. K., Jenal, N., Abdan, K., & Ali, A. (2012). Fatigue behavior of kenaf fibre reinforced epoxy composites. *Engineering Journal, 16*(5), 105–114.

Al-Oqla, F. M., & Salit, M. S. (2017). *Materials Selection for Natural Fiber Composites.* London: Woodhead Publishing.

Asgarinia, S., Viriyasuthee, C., Phillips, S., Dubé, M., Baets, J., Van Vuure, A., Verpoest, I., & Lessard, L. (2015). Tension—Tension fatigue behaviour of woven flax/epoxy composites. *Journal of Reinforced Plastics and Composites, 34*(11), 857–867.

Barter, S. A., Molent, L., & Wanhill, R. J. H. (2012). Typical fatigue-initiating discontinuities in metallic aircraft structures. *International Journal of Fatigue, 41*, 11–22.

Beevers, C. J. (1977). Fatigue crack growth characteristics at low stress intensities of metals and alloys. *Metal Science, 11*(8–9), 362–367.

Behzadi, S., Curtis, P. T., & Jones, F. R. (2009). Improving the prediction of tensile failure in unidirectional fibre composites by introducing matrix shear yielding. *Composites Science and Technology, 69*(14), 2421–2427.

Bensadoun, F. (2016). *In-service behaviour of flax fibre reinforced composites for high performance applications* [Doctoral Dissertation], Katholieke Universiteit Leuven, KU Leuven Research Repository, Leuven. https://lirias.kuleuven.be/1674573

Bensadoun, F., Vallons, K. A. M., Lessard, L. B., Verpoest, I., & Van Vuure, A. W. (2016). Fatigue behaviour assessment of flax—Epoxy composites. *Composites Part A: Applied Science and Manufacturing, 82*, 253–266.

Beyerlein, I. J., & Phoenix, S. L. (1997). Stress profiles and energy release rates around fiber breaks in a lamina with propagating zones of matrix yielding and debonding. *Composites Science and Technology, 57*(8), 869–885.

Blandford, R. S. (2001). *Characterization of Fatigue Crack Propagation in AA 7075-T651.* Starkville, MS: Mississippi State University.

Branco, R., Prates, P. A., Costa, J. D., Ferreira, J. A. M., Capela, C., & Berto, F. (2021). Notch fatigue analysis and crack initiation life estimation of maraging steel fabricated by laser beam powder bed fusion under multiaxial loading. *International Journal of Fatigue, 153*, 106468.

Broek, D. (1991). Fracture mechanics as an important tool in failure analysis. *Failure Analysis: Techniques and Applications, 33*–44.

Broek, D. (2012). *The Practical Use of Fracture Mechanics.* New York: Springer Science & Business Media.

Bruzzi, M. S., & McHugh, P. E. (2004). Micromechanical investigation of the fatigue crack growth behaviour of Al—SiC MMCs. *International Journal of Fatigue*, 26(8), 795–804.

Carlson, R. L., Kardomateas, G. A., & Craig, J. I. (2012). *Mechanics of Failure Mechanisms in Structures*. New York: Springer.

Case, S. W., & Reifsnider, K. L. (2003). *Fatigue of Composite Materials*. London: Elsevier.

Chell, G. G. (1979). A procedure for incorporating thermal and residual stresses into the concept of a failure assessment diagram. In *Elastic-plastic Fracture*. West Conshohocken, PA: ASTM International.

Collins, J. A. (1993). *Failure of Materials in Mechanical Design: Analysis, Prediction, Prevention*. Hoboken, NJ: John Wiley & Sons.

Curtin, W. A., & Takeda, N. (1998). Tensile strength of fiber-reinforced composites: I. Model and effects of local fiber geometry. *Journal of Composite Materials*, 32(22), 2042–2059.

de Vasconcellos, D. S., Touchard, F., & Chocinski-Arnault, L. (2014). Tension—Tension fatigue behaviour of woven hemp fibre reinforced epoxy composite: A multi-instrumented damage analysis. *International Journal of Fatigue*, 59, 159–169.

Ding, H.-Z., Biermann, H., & Hartmann, O. (2002). A low cycle fatigue model of a short-fibre reinforced 6061 aluminium alloy metal matrix composite. *Composites Science and Technology*, 62(16), 2189–2199.

Ding, H.-Z., Biermann, H., & Hartmann, O. (2003). Low cycle fatigue crack growth and life prediction of short-fibre reinforced aluminium matrix composites. *International Journal of Fatigue*, 25(3), 209–220.

Dobah, Y., Bourchak, M., Bezazi, A., Belaadi, A., & Scarpa, F. (2016). Multi-axial mechanical characterization of jute fiber/polyester composite materials. *Composites Part B: Engineering*, 90, 450–456.

Dowling, N. E. (1993). Mechanical behavior of metals-engineering methods for deformation. *Fracture, and Fatigue*. https://www.pearsonhighered.com/assets/preface/0/1/3/1/0131395068.pdf

El Messiry, M. (2017). *Natural Fiber Textile Composite Engineering*. London: CRC Press.

El Sawi, I., Fawaz, Z., Zitoune, R., & Bougherara, H. (2014). An investigation of the damage mechanisms and fatigue life diagrams of flax fiber-reinforced polymer laminates. *Journal of Materials Science*, 49(5), 2338–2346.

Ellyin, F., Carroll, M., Kujawski, D., & Chiu, A. S. (1997). The behavior of multidirectional filament wound fibreglass/epoxy tubulars under biaxial loading. *Composites Part A: Applied Science and Manufacturing*, 28(9–10), 781–790.

Ewing, J. A., & Humfrey, J. C. W. (1903). VI. The fracture of metals under repeated alternations of stress. *Philosophical Transactions of the Royal Society of London. Series A, Containing Papers of a Mathematical or Physical Character*, 200(321–330), 241–250.

Ezeh, O. H., & Susmel, L. (2018). On the fatigue strength of 3D-printed polylactide (PLA). *Procedia Structural Integrity*, 9, 29–36.

Ezeh, O. H., & Susmel, L. (2019). Fatigue strength of additively manufactured polylactide (PLA): effect of raster angle and non-zero mean stresses. *International Journal of Fatigue*, 126, 319–326.

Forrest, P. G. (2013). *Fatigue of Metals*. London: Elsevier.

Fotouh, A. (2014). *Characterization and modeling of natural-fibres-reinforced composites (moisture absorption kinetics, monotonic behaviour and cyclic behaviour)*. https://era.library.ualberta.ca/items/3439f845-2cd9-4958-ab12-c4007e8a652e/view/8ad4df13-ce47-47ff-a73f-f3125069bf88/Ahmed_Fotouh_PhD_201409.pdf

Fotouh, A., Wolodko, J. D., & Lipsett, M. G. (2014). Fatigue of natural fiber thermoplastic composites. *Composites Part B: Engineering*, 62, 175–182.

Gao, D. Y., Yao, W. X., Wen, W. D., & Huang, J. (2021). Equivalent spectral method to estimate the fatigue life of composite laminates under random vibration loadings. *Mechanics of Composite Materials*, *57*(1), 101–114.

Gassan, J. (2002). A study of fibre and interface parameters affecting the fatigue behaviour of natural fibre composites. *Composites Part A: Applied Science and Manufacturing*, *33*(3), 369–374.

Gokhale, N. S. (2008). *Practical Finite Element Analysis*. Pune: Finite to Infinite.

Harlow, D. G., & Phoenix, S. L. (1981). Probability distributions for the strength of composite materials II: A convergent sequence of tight bounds. *International Journal of Fracture*, *17*(6), 601–630.

Heinecke, F., & Willberg, C. (2019). Manufacturing-induced imperfections in composite parts manufactured via automated fiber placement. *Journal of Composites Science*, *3*(2), 56.

Hojo, T., Xu, Z., Yang, Y., & Hamada, H. (2014). Tensile properties of bamboo, jute and kenaf mat-reinforced composite. *Energy Procedia*, *56*, 72–79.

Hollaway, L. C. (2010). A review of the present and future utilisation of FRP composites in the civil infrastructure with reference to their important in-service properties. *Construction and Building Materials*, *24*(12), 2419–2445.

Ismail, A. E., & Aziz, M. A. C. A. (2015). Fatigue strength of woven kenaf fiber reinforced composites. *IOP Conference Series: Materials Science and Engineering*, *100*(1), 12037.

Katogi, H., Shimamura, Y., Tohgo, K., & Fujii, T. (2012). Fatigue behavior of unidirectional jute spun yarn reinforced PLA. *Advanced Composite Materials*, *21*(1), 1–10.

King, T. R., Blackketter, D. M., Walrath, D. E., & Adams, D. F. (1992). Micromechanics prediction of the shear strength of carbon fiber/epoxy matrix composites: The influence of the matrix and interface strengths. *Journal of Composite Materials*, *26*(4), 558–573.

Laird, C., & Smith, G. C. (1963). Initial stages of damage in high stress fatigue in some pure metals. *Philosophical Magazine*, *8*(95), 1945–1963.

Laws, N., & Dvorak, G. J. (1987). The effect of fiber breaks and aligned penny-shaped cracks on the stiffness and energy release rates in unidirectional composites. *International Journal of Solids and Structures*, *23*(9), 1269–1283.

Li, C., & Ellyin, F. (2000). A mesomechanical approach to inhomogeneous particulate composite undergoing localized damage: Part II—Theory and application. *International Journal of Solids and Structures*, *37*(10), 1389–1401.

Liang, S., Gning, P.-B., & Guillaumat, L. (2012). A comparative study of fatigue behaviour of flax/epoxy and glass/epoxy composites. *Composites Science and Technology*, *72*(5), 535–543.

Liang, S., Gning, P.-B., & Guillaumat, L. (2014). Properties evolution of flax/epoxy composites under fatigue loading. *International Journal of Fatigue*, *63*, 36–45.

Mandell, J. F., Huang, D. D., & McGarry, F. J. (1981). Fatigue of glass and carbon fiber reinforced engineering thermoplastics. *Polymer Composites*, *2*(3), 137–144.

McDowell, D. L. (1997). An engineering model for propagation of small cracks in fatigue. *Engineering Fracture Mechanics*, *56*(3), 357–377.

McDowell, D. L., Gall, K., Horstemeyer, M. F., & Fan, J. (2003). Microstructure-based fatigue modeling of cast A356-T6 alloy. *Engineering Fracture Mechanics*, *70*(1), 49–80.

Mir, A., Aribi, C., & Bezzazi, B. (2014). World Academy of Science, Engineering and Technology, *International Journal of Materials and Metallurgical Engineering*, *8*(2), 182–186.

Miraftab, M. (2009). *Fatigue Failure of Textile Fibres*. London: Elsevier.

Muthukumar, C. (2019). *Mechanical properties of fibre metal laminates reinforced with carbon, flax and sugar palm fibre-based composites* [Doctoral Dissertation], Universiti Putra Malaysia, Universiti Putra Malaysia Research Repository. http://psasir.upm.edu.my/id/eprint/77618/1/FK%202019%2019%20IR.pdf

Newman, Jr, J. C. (1998). The merging of fatigue and fracture mechanics concepts: A historical perspective. *Progress in Aerospace Sciences*, *34*(5–6), 347–390.

Nishida, S. (2014). *Failure Analysis in Engineering Applications*. London: Elsevier.

Orowan, E. (1939). Theory of the fatigue of metals. *Proceedings of the Royal Society of London. Series A. Mathematical and Physical Sciences*, *171*(944), 79–106.

Osgood, C. C. (2013). *Fatigue Design: International Series on the Strength and Fracture of Materials and Structures*. London: Elsevier.

Padmaraj, N. H., Chethan, K. N., & Onkar, A. (2017). Fatigue behaviour and life assessment of jute-epoxy composites under tension-tension loading. *IOP Conference Series: Materials Science and Engineering*, *225*(1), 12017.

Pickering, K. (2008). *Properties and Performance of Natural-fibre Composites*. London: Elsevier.

Pitt, R. E., & Phoenix, S. L. (1983). Probability distributions for the strength of composite materials IV: Localized load-sharing with tapering. *International Journal of Fracture*, *22*(4), 243–276.

Priyanka, P., Dixit, A., & Mali, H. S. (2017). High-strength hybrid textile composites with carbon, kevlar, and E-glass fibers for impact-resistant structures. A review. *Mechanics of Composite Materials*, *53*(5), 685–704. https://doi.org/10.1007/s11029-017-9696-2

Prajapati, H., Tevatia, A., & Dixit, A. (2021). Advances in natural fiber reinforced composites: A topical review. *Mechanics of Composite Materials, 58*(3), 319–354.

Roylance, D. (2001). *Introduction to Fracture Mechanics*. Cambridge, MA: Massachusetts Institute of Technology.

Sangrey, D. A., Henkel, D. J., & Esrig, M. I. (1969). The effective stress response of a saturated clay soil to repeated loading. *Canadian Geotechnical Journal*, *6*(3), 241–252.

Sclater, N., & Chironis, N. P. (2001). *Mechanisms and Mechanical Devices Sourcebook* (vol. 3). New York: McGraw-Hill.

Senalp, A. Z., Kayabasi, O., & Kurtaran, H. (2007). Static, dynamic and fatigue behavior of newly designed stem shapes for hip prosthesis using finite element analysis. *Materials & Design*, *28*(5), 1577–1583.

Shah, D. U. (2016). Damage in biocomposites: Stiffness evolution of aligned plant fibre composites during monotonic and cyclic fatigue loading. *Composites Part A: Applied Science and Manufacturing*, *83*, 160–168.

Shah, D. U., Schubel, P. J., Clifford, M. J., & Licence, P. (2013). Fatigue life evaluation of aligned plant fibre composites through S—N curves and constant-life diagrams. *Composites Science and Technology*, *74*, 139–149.

Shahzad, A., & Isaac, D. H. (2014). Fatigue properties of hemp and glass fiber composites. *Polymer Composites*, *35*(10), 1926–1934.

Sodoke, F. K., Toubal, L., & Laperrière, L. (2016). Hygrothermal effects on fatigue behavior of quasi-isotropic flax/epoxy composites using principal component analysis. *Journal of Materials Science*, *51*(24), 10793–10805.

Talreja, R., & Varna, J. (2015). *Modeling Damage, Fatigue and Failure of Composite Materials*. London: Elsevier.

Tevatia, A., & Srivastava, S. K. (2015). Modified shear lag theory based fatigue crack growth life prediction model for short-fiber reinforced metal matrix composites. *International Journal of Fatigue*, *70*, 123–129.

Tevatia, A., & Srivastava, S. K. (2018). The energy-based multistage fatigue crack growth life prediction model for DRMMC s. *Fatigue & Fracture of Engineering Materials & Structures*, *41*(12), 2530–2540.

Towo, A. N., & Ansell, M. P. (2008a). Fatigue evaluation and dynamic mechanical thermal analysis of sisal fibre-thermosetting resin composites. *Composites Science and Technology*, *68*(3–4), 925–932. https://doi.org/10.1016/j.compscitech.2007.08.022

Towo, A. N., & Ansell, M. P. (2008b). Fatigue of sisal fibre reinforced composites: Constant-life diagrams and hysteresis loop capture. *Composites Science and Technology, 68*(3–4), 915–924.

Ueki, Y., Lilholt, H., & Madsen, B. (2015). Fatigue behaviour of uni-directional flax fibre/epoxy composites. In *Proceedings of the 20th International Conference on Composite Materials ICCM20 Secretariat.*

Wang, A. S. D., Chou, P. C., & Lei, S. C. (1984). A stochastic model for the growth of matrix cracks in composite laminates. *Journal of Composite Materials, 18*(3), 239–254.

Wasage, T., Statsna, J., & Zanzotto, L. (2010). Repeated loading and unloading tests of asphalt binders and mixes. *Road Materials and Pavement Design, 11*(3), 725–744.

Wu, S. C., Zhang, S. Q., Xu, Z. W., Kang, G. Z., & Cai, L. X. (2016). Cyclic plastic strain based damage tolerance for railway axles in China. *International Journal of Fatigue, 93*, 64–70.

Xue, S., Shen, R., Chen, W., & Miao, R. (2020). Corrosion fatigue failure analysis and service life prediction of high strength steel wire. *Engineering Failure Analysis, 110*, 104440.

Yang, J. N., & Jones, D. L. (1980). Effect of load sequence on the statistical fatigue of composites. *AIAA Journal, 18*(12), 1525–1531.

Yuanjian, T., & Isaac, D. H. (2007). Impact and fatigue behaviour of hemp fibre composites. *Composites Science and Technology, 67*(15–16), 3300–3307.

Zhao, X., Wang, X., Wu, Z., Keller, T., & Vassilopoulos, A. P. (2018). Effect of stress ratios on tension—Tension fatigue behavior and micro-damage evolution of basalt fiber-reinforced epoxy polymer composites. *Journal of Materials Science, 53*(13), 9545–9556.

Zhou, J. Z., Huang, S., Sheng, J., Lu, J. Z., Wang, C. D., Chen, K. M., Ruan, H. Y., & Chen, H. S. (2012). Effect of repeated impacts on mechanical properties and fatigue fracture morphologies of 6061-T6 aluminum subject to laser peening. *Materials Science and Engineering: A, 539*, 360–368.

Zimnicwska, M., Wladyka-Przybylak, M., & Mankowski, J. (2011). Cellulosic bast fibers, their structure and properties suitable for composite applications. In *Cellulose Fibers: Bio-and Nano-polymer Composites* (pp. 97–119). New York: Springer.

8 Finite Element Method Formidable in Damage Modelling of Green Composites

Mukesh Kumar, Abhishek Tevatia, and Anurag Dixit

CONTENTS

8.1 INTRODUCTION

The demand of natural fibers (NFs) as reinforcements in polymer composites is increasing due to their recyclability (Yan et al., 2014), eco-friendliness (Mohammed et al., 2015), and biodegradability (Al-Oqla & Sapuan, 2014) as compared with synthetic fibers. Plant-based NFs have the capability to partially replace synthetic fibers such as glass or carbon due to their acceptable range of characteristics (Alves et al., 2010; Yan et al., 2014). NFs possess some important characteristics such as stiffness (Aji et al., 2013; Dittenber & Gangarao, 2012; Wecławski et al., 2014), toughness, specific strength, and flexibility (Kiruthika, 2017). Furthermore, NFs are abundant (Xiong, Shen et al., 2018), renewable (Hughes et al., 2007), and sustainable (Zhang et al., 2019). The significant advantages of NFs include lower price, lower mass density, good acoustic and thermal insulation, less equipment abrasion, environmental averseness, acceptable specific mechanical properties, noticeable energy recovery, and less respiratory and skin irritation (Gholampour & Ozbakkaloglu, 2020; Mishra & Biswas, 2013; Sgriccia et al., 2008). NFs possess some constraints, which include lower mechanical strength, large variability, and inferior moisture resistance, and lower impact resistance compared to man-made fibers (Caprino et al., 2015; Ravandi et al., 2017). The primary

DOI: 10.1201/9781003272625-8

applications of NFCs are in textile (Miao & Shan, 2011; Xue et al., 2011), automotive (Fu et al., 2012; Kong et al., 2016), and construction industries (Singh & Gupta, 2005; Wang et al., 2016), while some limited applications of NFCs are confined to secondary structures such as aerospace industry (Scarponi, 2015; Scarponi & Messano, 2015) and non-structural areas (Dicker et al., 2014; Mohammed et al., 2015; Naveen et al., 2018; Singh & Gupta, 2005).

In few applications such as automotive (Fu et al., 2012; Kong et al., 2016), aerospace (Scarponi, 2015; Scarponi & Messano, 2015), and structural (Singh & Gupta, 2005), etc., damages can cause sudden fracture and even failure in NFCs structures. Consequently, the evaluation of damage in NFCs' structural components is quite significant so that its substitution can be planned well in advance to avoid any catastrophic failure. Also, with increasing demand of NFCs, it has become a necessity to design and develop NFCs for optimal performance. Researchers are adopting computer aided numerical techniques to predict the thermal as well as mechanical related properties of NFCs. It's not always possible to attain exact solutions of real-life problems such as complex and irregular geometry, nonlinear material properties, and nonuniform loading conditions. In such situations, however, approximate solutions obtained by computer aided numerical techniques are the best bets. A variety of numerical techniques, such as finite volume method (FVM), finite difference method (FDM), and FEM are available for solving complex problems. Amongst these available techniques, FEM is the more popular and commonly used in industry and academics for modelling of NFCs because it can deal with any complex geometry and irregular shape, material anisotropy and inhomogeneity, nonuniform loading, and any boundary conditions (Kong et al., 2016; Mohammed Ameen, 2001; Reddy, 2019; Saavedra Flores & Haldar, 2016).

The thermal performance and mechanical strength of NFCs can be predicted by FEM using the parameters like thermal conductivity and density, Poisson's ratio, and elastic constants. The damage failure can be predicted with FEM using the loading conditions and material properties of the matrix, fiber, and their interface. However, utilizing and developing the right failure criteria are the key factors of finite element (FE) modelling. There are various types of FEM models available for analysis of failure of NFCs at different scales. The outlines of this chapter are as follows: (i) introducing the green composites, (ii) steps involved in FEM analysis in damage, and (iii) FE models on failure and damage of NFCs.

8.2   GREEN COMPOSITES

Fibers have a structure similar to a hair-like structure and may be turned into either thread, filament, or rope for various uses (Chandramohan & Marimuthu, 2011; Nassar et al., 2017). Fibers are classified as natural and synthetic. Further NFs are classified as mineral, plant, and animal type. Figures 8.1 and 8.2 are flowcharts to classify fibers (in general) and plant origin (cellulosic/lignocellulose) fibers, respectively. The role of a matrix in fiber reinforced composites (FRCs) is to grip the fibers together and increase the durability of the composites, surface appearance, and environmental tolerance (Campilho, 2015).

The mechanical properties of green composites were found to be lower as compared to that of man-made fiber composites. However, these properties can further be

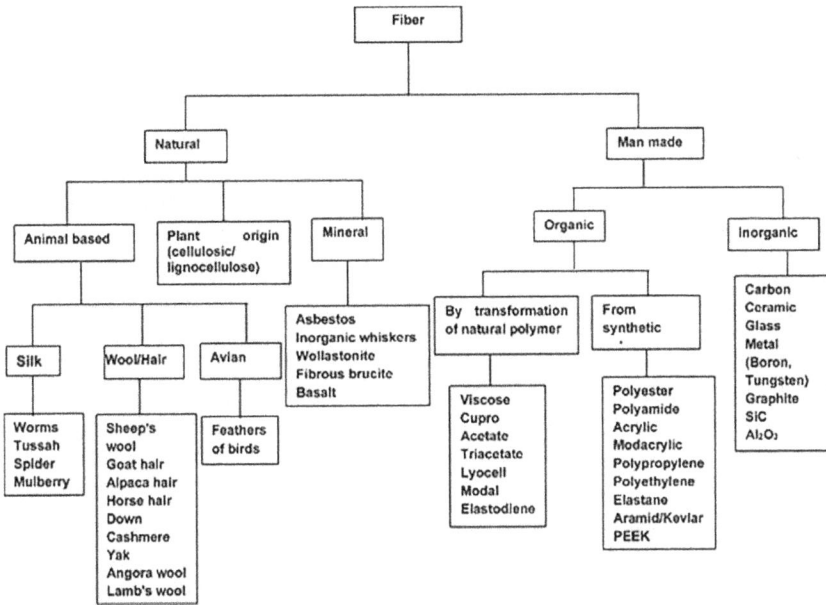

FIGURE 8.1 Classification of fibers.

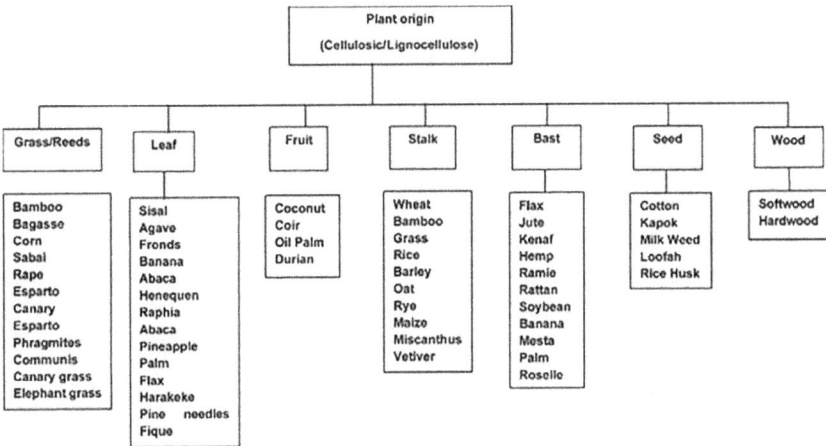

FIGURE 8.2 Classification of plant origin (cellulosic/lignocellulose) fibers.

enhanced by proper chemical and physical treatments of the NFs and matrix (Prajapati et al., 2021). The tensile property of green composites is mainly affected by the fiber volume fraction and interfacial bonding between the matrix-fibers. Generally, the tensile strength of NFCs increases with increasing fiber volume fraction up to the optimum value. However, when the fiber volume fraction exceeds the optimal value then there is a brittle failure in fiber because of low tensile strength (Gholampour & Ozbak-kaloglu, 2020). Flexural properties are mainly affected by the moment of inertia and the modulus of the composite materials. The flexural strength of green composites increases with increasing fiber content up to the optimal amount. However, there is a decrement in flexural strength with further increment in fiber content (Faruk et al., 2012). The compressive strength of NFCs is mainly affected by fiber volume fraction and fiber architecture (Wecławski et al., 2014). Compressive strength is increased with increase in fiber volume fraction up to the optimal value, which also results in a smaller number of voids. However, there is a decrement in compressive strength when fiber volume fraction exceeds the optimum value. The fracture toughness i.e. resistance to crack propagation is mainly affected by fiber volume fraction, the interfacial bonding strength between matrix-fibers, and the inherent properties of matrix (Faruk et al., 2012; Prajapati et al., 2021). The impact strength of NFCs is mainly affected by the level of bonding between fibers and matrix, which plays a significant role during the service life of NFCs (Faruk et al., 2012).

8.3   DAMAGE MECHANICS IN COMPOSITE

Composite material failures are complex mechanisms because of their manufacturing processes and heterogeneous and anisotropic nature, which include fiber debonding and pull-out, fiber fracture, interface failure, buckling and delamination, fiber micro-buckling, and matrix failure as shown in Figure 8.3. Analytical, numerical, and experimental techniques were developed to predict these failures by several groups (Babaei & Farrokhabadi, 2020; Guillebaud-Bonnafous et al., 2012; Hanipah et al., 2020; Kern et al., 2016; Khaldi et al., 2016; Mahboob et al., 2017; Panamoottil et al., 2016; Shokrian et al., 2016; Távara et al., 2019; Tevatia & Srivastava, 2015; Wang et al., 2015; Zhang et al., 2020). Cracks/damaged areas can be formed due to these failure mechanisms and may be active in loading situations. These local cracks/damage interact with each other and finally may lead to failure at the composite level (Talreja, 1985).

The simplest failure in a composite is fiber failure, which can be easily recognized and quantified, and happens when external stresses are applied to composites. During this failure, fracture takes place in fiber. In the compression mode, this failure takes place because of the formation of kink bands and micro-buckling. On contrary, in the tension mode, this failure takes place due to build-up of individual fiber failures inside plies. Furthermore, this turns crucial when there are not enough fibers to bear applied loads.

Delamination is observed as the common type of failure observed in laminated composite, which is separation between internal layers because of their moderately weak interlaminar strengths (Zou et al., 2003). This failure may arise due to a manufacturing defect (Camanho et al., 2003) such as broken fibers, imperfect bonding and crack in matrix materials, and in service loading such as fatigue (Mahboob & Bougherara, 2018) and impact (Camanho et al., 2003; Zou et al., 2003). Nonetheless, this failure can cause a major reduction in the compressive strength of a composite structure (Camanho et al., 2003) and the ultimate result of this may lead to significant structural damage, especially in the compression mode (Orifici et al., 2008).

Both pull-out and debonding failures are observed at the microscopic scale. Debonding is defined as full or partial damage of connection between matrix and fibers because

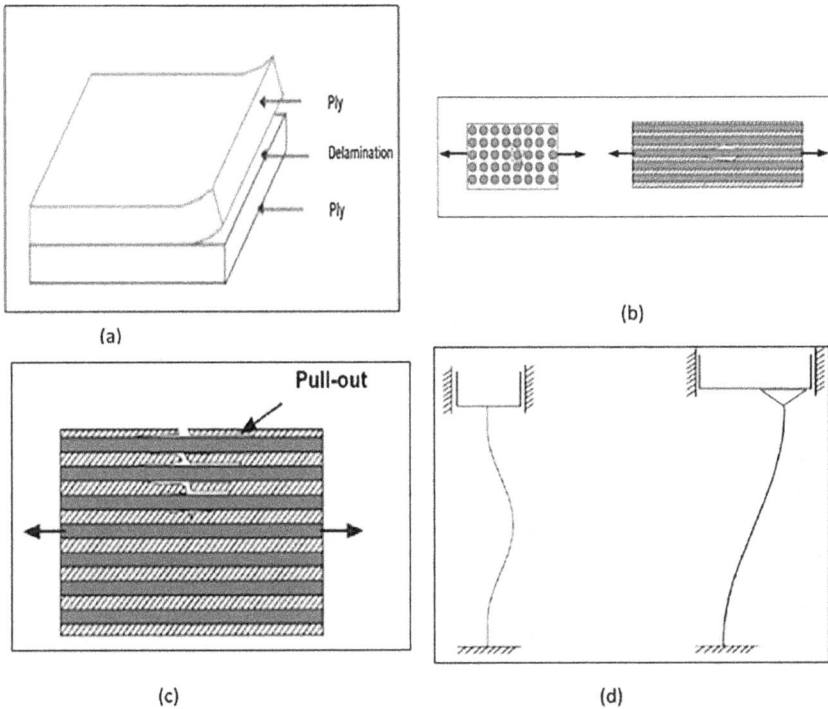

FIGURE 8.3   Various failure mechanisms in NFCs: (a) delamination, (b) matrix-fibers debonding, (c) matrix-fibers pull-out, and (d) fiber micro-buckling and fiber kinking.

of external applied load. It takes place in composites when applied load exceeds the interface shear strength. Pull-out needs debonding as an initial damage state, followed by slipping of the fibers with respect to the matrix, where the communication between separated fibers region and matrix is purely by friction. Micro-buckling is observed in composites when they are compressed towards fibers. Generally, most FRCs having lower compressive strength than their tensile strength due to micro-buckling, and hence compressive strength is selected as a design criterion in many applications (Fleck & Budiansky, 1991). The last stage of micro-buckling is fiber kinking. Moreover, this failure mode has been identified as the dominant failure mode restricting the compressive strength of that composite.

Intralaminar form of damages come under matrix failures, which include void or cracks among fibers within a single lamina of composite. This is a complex phenomena, in which matrix cracks initiate typically at fiber-matrix interfaces and finally are responsible for failure across a fracture plane.

8.4   FINITE ELEMENT ANALYSIS (FEA) OF NFCS

FEA is a simulation and modelling tool and is applied where close form solution is impossible, hence it caters to approximate solutions. It appears that FEA can handle

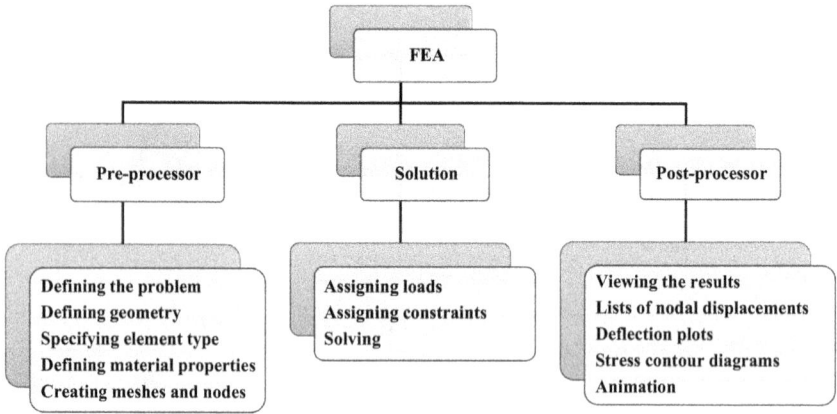

FIGURE 8.4 The general steps of FEA.

almost all engineering science real life scenarios with any complex geometry and irregular shapes, nonlinear material properties, and nonuniform loading conditions (Dixit & Harlal, 2013; Fattahi et al., 2019; Naveen et al., 2018). To ensure optimized and precise results, researchers perform several iterations to improve its lifetime and reduce the product development down time. The general steps of the FEA are shown in Figure 8.4.

8.4.1 GENERAL STEPS OF FEA

The first and most significant step of FEA is pre-processing. The accuracy of results and quality of the simulation mainly depend on pre-processing. 1-dimensional (1D), 2-dimensional (2D), or 3-dimensional (3D) modelling of the given problem geometry, material definition, element, and meshing are performed in pre-processing. Various commercial FEA software programs are available to perform pre-processing. Modelling the complex geometry such as analysis of textile composites, aircraft fuselage design, and car body design requires separate dedicated software for modelling and meshing. Presently, TexGen and HYPERMESH are successfully used for modelling of textile composites and meshing of complex geometry respectively (Naveen et al., 2018; Sharma et al., 2019).

Imposition of appropriate boundary conditions (BC's) such as thermal, electrical, magnetic, and structural BC's, and application of loads like electrical, thermal, magnetic, and structural loads are incorporated in the solution phase, which depends on considered problem types. A dedicated solver is used to solve elemental and global equations, as represented by equations (1) and (2) respectively (Reddy, 2019).

$$[k]_{dof.nne \times dof.nne} \{u\}_{dof.nne \times 1} = \{f\}_{dof.nne \times 1} \tag{1}$$

$$[GK]_{dof.nnm \times dof.nnm} \{U\}_{dof.nnm \times 1GF} = \{GF\}_{dof.nnm \times 1} \tag{2}$$

Where, $[k]$, $[GK]$, $\{u\}$, $\{U\}$, $\{f\}$, and $\{GF\}$ are elemental stiffness matrix, global stiffness matrix, elemental displacement vector, global displacement vector, elemental

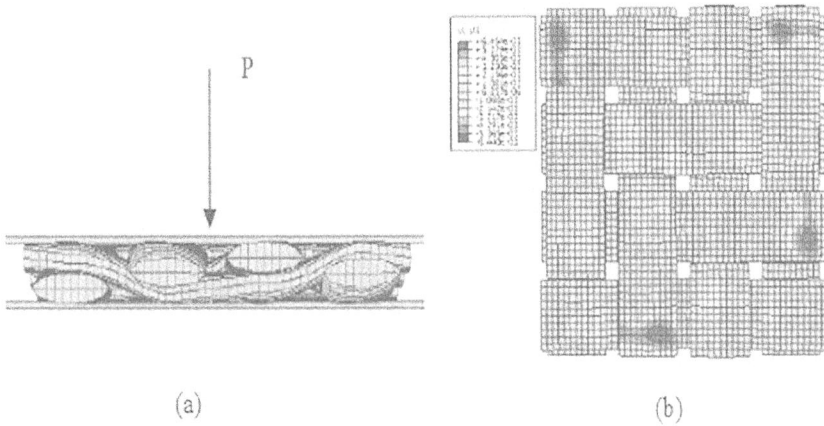

FIGURE 8.5   (a) Pre-processing of twill weave under compression and (b) he post-processing results of twill weave under compression.

force vector, and global force vector respectively, *nne* and *nnm* are number of nodes in an element and mesh respectively, and *dof* is degree of freedom per node.

Improper meshing, loading conditions, and boundary conditions can lead to errors in solving the FE equations. In general, the accuracy of the solution is a function of the order and size of elements. The accuracy can be increased by either increasing the order of an element or reducing the size of element, i.e., total number of elements in mesh is increased, which also leads to increase computational time, i.e., computational cost. Hence, FE user has to be careful about the number of elements in the mesh, so that accuracy of results increases with decrease in overall computational cost. For this purpose, convergence test is carried out by the FE user, which decides the optimal number of elements in the mesh for the given analysis type (Naveen et al., 2018).

The prime objective of the post-processing is to demonstrate the FEA results, which include nodal stress, strain, and displacement. Results in the form of animations, graphs, plots, and tables are generated in the post-processor step. Figure 8.5 shows the pre-processing and post-processing results of a twill weave under compression; maximum displacements were found over cross over of yarns.

8.4.2   REPRESENTATIVE VOLUME ELEMENT (RVE) MODELS OF NFCS

FEA is broadly accepted by researchers to model the NFCs (Assarar et al., 2015; Kern et al., 2016; Modniks & Andersons, 2010; Rafiquzzaman et al., 2016; Reis et al., 2007; Sliseris et al., 2016). FEA estimates the consequence of orientation, aspect ratio, volume fraction of natural fibers, and reinforcement and predicts the thermal and mechanical behaviour of NFCs (Naveen et al., 2018; Xiong, Shen et al., 2018). As discussed previously natural fiber is made of hemicellulose, pectin, lignin, and cellulose (Gholampour & Ozbakkaloglu, 2020). The chemical composition and microstructure of natural fibers decide the macroscale properties of natural fibers (Saavedra Flores & Haldar, 2016). It is compulsory to use a homogenization-based multi-scale computational

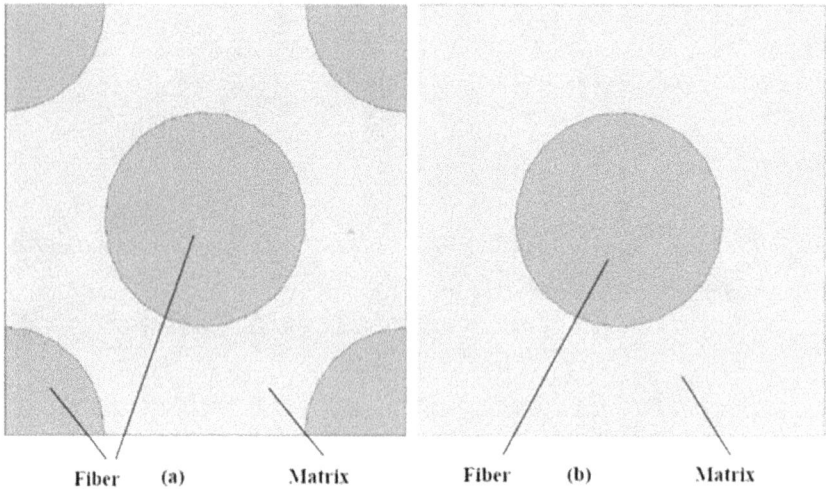

FIGURE 8.6   RVE for (a) hexagonal array distribution (b) square array distribution.

technique to efficiently investigate the connection between macrostructural and micro-structural (Sun & Vaidya, 1996). The representative volume element is one of the most accepted homogenization-based multiscale FE models that represent the appropriate characteristics of NFCs in a uniform microstructure (Sun & Vaidya, 1996).

Because of the huge number of elements, random distribution of fiber across the cross-section, and modelling complexity, it is difficult to model and analyze the whole composite structure (Sharma et al., 2019). For simplicity, micromechanical models opt for the representation of a small part of the laminate composite. Furthermore, it is assumed that the whole composite structure can be represented by several repetitions of this volume in all directions, which is called a unit cell or RVE. Hence, RVE represents the smallest volume, which has similar fiber volume fraction and elastic constants as the composite (Drugan & Willis, 1996). The hexagonal and the square arrays are frequently used for periodic fiber sequences (Figure 8.6). Currently, computational tools like material designer in ANSYS and EasyPBC in ABAQUS are efficiently used to simulate the unit cell of NFCs. The volume fraction, material's properties, and fiber size are input to these tools. Thereby these tools have the ability to automatically define the dimensions of the corresponding unit cell, appropriate mesh type, and size and ultimately solve the unit cell model. Homogeneous traction BCs, homogeneous displacement BCs, and periodic BCs are preferred boundary conditions for RVE or unit cell analysis among these three boundary conditions. PBC is considered the best boundary condition for RVE analysis, numerically proved and theoretically justified (Rahman, 2017; Yu, 2016).

A several FEM models are available in literature to estimate the mechanical prop-erties of NFCs. The list includes 2-dimensional macroscale model, orientation aver-aging approach (OAA), direct 3-dimensional RVE model, Multiscale RVE model and 3-dimensional macroscale model. A 2D macroscale FEM model was developed to study the damping behaviour and bending stiffness of hybrid laminated composites with different stacking sequences (Assarar et al., 2015). Integrated energy approach was opted for analyzing the structural related damping behaviour of the composites. However, it may be concluded that the flax fiber position plays an important role on the damping behaviour and bending stiffness of hybrid laminated composite (Assarar et al., 2015).

In another study, a FEM orientation averaging approach was developed for the study of mechanical behaviour of NFCs, a simplified and less time-consuming computing. Modniks and Andersons (Modniks & Andersons, 2010) estimated the elastic properties and stiffness of unit cell of flax FRCs by employing the OAA under axial tension, shear, transverse tension, and equibiaxial tension pure shear loading conditions. Furthermore, they (Modniks & Andersons, 2013) estimated the non-linear deformation of flax FRCs by using the OAA under transverse tension, transverse compression, axial tension, equibiaxial tension pure shear, and shear loading conditions. They concluded that stress-strain diagram predicted from the OAA showed good conformity with experimental results up to 1.5% strain for fiber volume fractions equal to 0.29, 0.2, and 0.13.

Kern et al. (Kern et al., 2016) proposed a 3-dimensional RVE model to investigate the effect of void and fiber on the tensile behaviour of short fiber reinforced composite (SFRCs). Loisirs et al. (Sliseris et al., 2016) proposed a 3D RVE model to study the effect of fiber defects, fiber bundles, and different aspect ratios of fiber on the tensile behaviour of SFRCs. In this model, linear isotropic elastic, brittle, nonlinear plastic, and brittle materials were used to model the fiber, fiber bundles, matrix, and fiber defects respectively. Further, they concluded that deformation started at the vicinity of the defects and at the fiber endings. Hosseini et al. (Hosseini et al., 2015) also proposed a 3D RVE model to investigate viscoelastic characteristics of NFCs by assuming perfect bonding between matrix and fiber, and a cylindrical cross-sectional area of the fiber.

Reis et al. (Reis et al., 2007) proposed a 3-dimensional FE macroscale model for investigating flexural properties, and fatigue behaviour in the three-point bending test of hemp/glass hybrid eliminated composites. Rafiquzzaman et al. (Rafiquzzaman et al., 2016) also developed a 3D macroscale FE model to study the mechanical behaviour of jute-glass FRCs. They concluded that by incorporating the optimal quantity of jute fiber (30%), the overall strength of glass FRCs can be improved, also the overall price can be decreased by more than 30%. Zhong et al. (Zhong et al., 2017) developed an FEM multi-scale RVE model to study the mechanical behaviour of flax FRCs. This model predicted the bending strength, major failure modes, and tensile strength of flax FRCs.

Xiong et al. (Xiong, Hua et al., 2018) proposed a 3-dimensional microscale unit cell model to study the effect of yarn twist angle on the mechanical behaviour of plain weave NFCs. The multi-scale constitutive modelling was performed in two parts: First OAA was used to find micromechanical properties of the twisted yarn and then results of part one were utilized into a mesoscale RAV model to predict elastic properties of composites. They suggested that the yarn twist angle significantly affects the elastic behaviour of NFCs.

8.4.3   FINITE ELEMENT-BASED DAMAGE MODELS FOR NFCs

Currently, FEA of the failures of NFCs is mostly limited on matrix-fiber interface debonding as the interface shear strength of NFCs is usually inferior, and reduces sharply under moist conditions (Chen et al., 2009; D. Zhang et al., 2014). Khaldi et al. (Khaldi et al., 2016) proposed a FE model for alfa NFCs to examine the initiation of crack at micro level and its propagation at fiber-matrix interface. In this model, the alfa fiber and the matrix were modelled as an anisotropic elastic material and a viscoelastic material respectively. They integrated thermodynamics to evaluate initiation and propagation of crack in composites. They implemented a small debonding area at fiber-matrix interphase to describe the defect at interphase. Their results suggested that a micro crack originated in the matrix and propagated perpendicular to the applied load when a load is applied transverse to the fiber axis. They additionally observed that propagation of matrix edge cracks more readily than interior matrix cracks. Furthermore, they also found that the energy release rate decreases in near interface, and this becomes higher due to generation of interfacial debonding areas because of fiber wettability.

Wang et al. (W. Wang et al., 2015) developed Forming Limit Curve (FLC) for flax FRCS and incorporated in 2D FEM with the maximum strain failure theory. They identified that both maximum strain failure criteria and new FLC can predict failure regions in all specimens. They also inferred that the accuracy of FEA prediction considerably improved by the incorporation of novel FLC as compared to the maximum strain failure criteria specially for 0°/90° specimens. Keck and Fulland (Keck & Fulland, 2016) developed a 3D FEM for fatigue analysis of flex FRCs to investigate the stress intensity factor evolution along the growing crack and determined the fatigue crack growth rate curve by taking into account the fiber volume fraction and fiber direction respectively. In this present model, composites were modelled homogeneous and orthotropic materials. They opted for linear elastic fracture mechanics for the analysis. They observed that both fiber orientation as well as fiber volume fraction influenced the crack growth rate curve. The maximum fracture toughness was obtained when the load was perpendicular to the fiber orientation.

An anisotropic continuum damage (CDM) model was proposed by Panamoottil et al. (Panamoottil et al., 2016) to examine the stress-strain behaviour and failure strength behaviour of the flax fiber polypropylene composites under uniaxial tension. The stress-strain curve and failure strength obtained from FE simulation were close to the experimental value with an error of 3.1%. Singh et al. (2020) developed a FE model to investigate the interfacial behaviour during pull-out of a single fiber in FRCs by using the *cohesive zone modelling* method. In this model hexahedron elements (C3D8I) were used to model matrix and fiber. To study interface strength, the sensitivity of interface stiffness, fiber length, shear stress distribution across the matrix-fiber interface zone, and friction coefficient, a 2D axisymmetric model was incorporated. They suggested that enough distribution of the sizing on the fiber resulted in three times increments in necessary force to debond a single fiber from the matrix. They also suggested that cohesive strength and fiber length played important roles in debonding.

Hanipah et al. (Hanipah et al., 2020) developed a 3-dimensional FE micromechanical model of oil palm fiber polyethylene composites to investigate the pull-out and the consequence of silica bodies on the function of composites. They concluded that the adding of silica bodies enhances the performance of the oil palm fiber polyethylene composites. Kern et al. (Kern et al., 2016) proposed a 3D FEM to study the mechanical behaviour of the NFCs with microcellular voids. They concluded that composites having microcellular voids possess a short fracture path as compared to solid composites due to development of stress concentration in the matrix owing to micro voids. They also suggested that fracture surfaces and fracture paths predicted from the simulation model more converged to the experimental values.

Sun et al. (X. Sun et al., 2019) developed a multiscale FE model to perform three-point bending tests and investigated the flexural behaviour on precast notched beams and the effect of various basalt fibers doses on fracture resistance for Basalt FRCs. They suggested that there could be an increment in fracture energy and peak load with an increase in basalt fiber dose.

Although fibers are the primary load-carrying components in composites, the interface regions of fiber-matrix recognized as the key factor in the failure of composites. Therefore, it is equally important to study the interfacial behaviour. Generally, for FEM of interface damage of FRCs three methods are often needed to study which include (i) an interphase or perfect bonding is used for boundary nodes near the interface of fiber-matrix (Heydari-Meybodi et al., 2015), (ii) to model the mechanical behaviour on matrix-fiber interphase frictional contact is applied (Guillebaud-Bonnafous et al., 2012; Zhi et al., 2017), and (iii) with a constitutive law the cohesive or spring elements are exploited to describe the interphase (Shokrian et al., 2016; Yang et al., 2012).

Guillebaud-Bonnafous et al. (Guillebaud-Bonnafous et al., 2012) proposed a 2-dimensional microscale RVE FE model of fragmentation tests to examine the properties related to adhesion of composites reinforced with hemp yarn FRCs. They considered two dissimilar contact interactions at interphase: (i) frictional contact and (ii) perfect bonding. They modelled the contact behaviour of the frictional contact using coulomb's friction mode with different friction coefficients. They observed that fragmentation patterns, procured from frictional contact models, are best fitted with coefficient friction of 0.7. They also observed good conformity between their experimental and simulated results in terms of isochromatic patterns and interfacial shear strength.

Alam et al. (Alam et al., 2015) proposed a plane-stress FEM model for NFCs by rearranging the 2-dimensional topographical profiles from contour-mapping of coating profiles as well as from SEM images of biomimetically mineralized fibers. In this simulation, evaluation of shear strain was performed in the rubber matrix near the surface of different $CaCO_3$/amino acid coatings. However, it was inferred that CaCO3/β-alanine and CaCO3/glycine coatings at higher shear strain points showed lower peak frequency as compared to that of CaCO3/L-lysine coatings. It showed more interlocking points and less separation chance of the $CaCO_3$/L-lysine coatings from the fiber surfaces.

8.5 CONCLUDING REMARKS

This chapter briefly describes major failure and damage related to modelling strategies for NFRCs. The FE modelling facilitates the virtual experiments of the actual complex phenomenon to be conducted and viewed in graphics user interface environment with ease and enables one to narrow down the number of complicated and exhaustive experimental stages. Representative volume element (RVE)/Unit cell and orientation averaging approach emerge to be the effective methods for predicting the mechanical behaviour of NFRCs.

Development and selection of appropriate failure criteria in conjunction with consistent material properties of fiber, matrix, and adequate loading condition are recognized as the key factors for FEM and the failure behaviour of NFRCs. The prime factors responsible to model the fiber/matrix interface damage in FEM simulations of NFRCs can be elaborated as frictional contact, perfect bonding, and cohesive elements. Unit cell/RVE method is found to be the most significant homogenization based constitutive technique in FE modelling for the evaluation of impact of microstructures on the mechanical and thermal properties of NFCs.

To lower the experimental cost and time in designing an appropriate and accurate FEM, analysis is required. The challenges involved in the aforementioned failure models are: defining the appropriate three-dimensional geometric model, finding the properties of interfacial adhesion, and defining the correct fiber-matrix interfaces. Though a lot of research has been done on FEM, the progressive damage analysis and mechanics of NFCs remains a prominent area of further research.

REFERENCES

Aji, I. S., Zainudin, E. S., Abdan, K., Sapuan, S. M., & Khairul, M. D. (2013). Mechanical properties and water absorption behavior of hybridized kenaf/pineapple leaf fibre-reinforced high-density polyethylene composite. *Journal of Composite Materials, 47*(8), 979–990. https://doi.org/10.1177/0021998312444147

Alam, P., Fagerlund, P., Hägerstrand, P., Töyrylä, J., Amini, S., Tadayon, M., Miserez, A., Kumar, V., Pahlevan, M., & Toivakka, M. (2015). L-Lysine templated CaCO3 precipitated to flax develops flowery crystal structures that improve the mechanical properties of natural fibre reinforced composites. *Composites Part A: Applied Science and Manufacturing, 75*, 84–88. https://doi.org/10.1016/j.compositesa.2015.04.016

Al-Oqla, F. M., & Sapuan, S. M. (2014). Natural fiber reinforced polymer composites in industrial applications: Feasibility of date palm fibers for sustainable automotive industry. *Journal of Cleaner Production, 66*, 347–354. https://doi.org/10.1016/j.jclepro.2013.10.050

Alves, C., Ferrão, P. M. C., Silva, A. J., Reis, L. G., Freitas, M., Rodrigues, L. B., & Alves, D. E. (2010). Ecodesign of automotive components making use of natural jute fiber composites. *Journal of Cleaner Production, 18*(4), 313–327. https://doi.org/10.1016/j.jclepro.2009.10.022

Assarar, M., Zouari, W., Sabhi, H., Ayad, R., & Berthelot, J. M. (2015). Evaluation of the damping of hybrid carbon-flax reinforced composites. *Composite Structures, 132*, 148–154. https://doi.org/10.1016/j.compstruct.2015.05.016

Babaei, R., & Farrokhabadi, A. (2020). Prediction of debonding growth in two-dimensional RVEs using an extended interface element based on continuum damage mechanics concept. *Composite Structures, 238*, 111981.

Camanho, P. P., Dávila, C. G., & De Moura, M. F. (2003). Numerical simulation of mixed-mode progressive delamination in composite materials. *Journal of Composite Materials, 37*(16), 1415–1438. https://doi.org/10.1177/0021998303034505

Campilho, R. D. S. G. (2015). Natural fiber composites. In *Natural Fiber Composites*. London: CRC Press. https://doi.org/10.1201/b19062

Caprino, G., Carrino, L., Durante, M., Langella, A., & Lopresto, V. (2015). Low impact behaviour of hemp fibre reinforced epoxy composites. *Composite Structures, 133*, 892–901. https://doi.org/10.1016/j.compstruct.2015.08.029

Chandramohan, D., & Marimuthu. (2011). A review on natural fibers. *International Journal of Research and Reviews in Applied Sciences, 8*(2), 194–206.

Chen, H., Miao, M., & Ding, X. (2009). Influence of moisture absorption on the interfacial strength of bamboo/vinyl ester composites. *Composites Part A: Applied Science and*

Manufacturing, 40(12), 2013–2019. https://doi.org/10.1016/j.compositesa.2009.09.003

Dicker, M. P. M., Duckworth, P. F., Baker, A. B., Francois, G., Hazzard, M. K., & Weaver, P. M. (2014). Green composites: A review of material attributes and complementary applications. *Composites Part A: Applied Science and Manufacturing, 56*, 280–289.

Dittenber, D. B., & Gangarao, H. V. S. (2012). Critical review of recent publications on use of natural composites in infrastructure. *Composites Part A: Applied Science and Manufacturing, 43*(8), 1419–1429. https://doi.org/10.1016/j.compositesa.2011.11.019

Dixit, A., & Harlal, S. M. (2013). Modeling techniques for predicting. *Mechanics of Composite Materials, 49*(1), 1–20.

Drugan, W. J., & Willis, J. R. (1996). A micromechanics-based nonlocal constitutive equation and estimates of representative volume element size for elastic composites. *Journal of the Mechanics and Physics of Solids, 44*(4), 497–524. https://doi.org/10.1016/0022-5096(96)00007-5

Faruk, O., Bledzki, A. K., Fink, H. P., & Sain, M. (2012). Biocomposites reinforced with natural fibers: 2000–2010. *Progress in Polymer Science, 37*(11), 1552–1596. https://doi.org/10.1016/j.progpolymsci.2012.04.003

Fattahi, A. M., Safaei, B., & Moaddab, E. (2019). The application of nonlocal elasticity to determine vibrational behavior of FG nanoplates. *Steel and Composite Structures, 32*(2), 281–292. https://doi.org/10.12989/scs.2019.32.2.281

Fleck, N. A., & Budiansky, B. (1991). Compressive failure of fibre composites due to microbuckling. In *Inelastic Deformation of Composite Materials* (pp. 235–273). New York: Springer.

Fu, Z., Suo, B., Yun, R., Lu, Y., Wang, H., Qi, S., Jiang, S., Lu, Y., & Matejka, V. (2012). Development of eco-friendly brake friction composites containing flax fibers. *Journal of Reinforced Plastics and Composites, 31*(10), 681–689. https://doi.org/10.1177/0731684412442258

Gholampour, A., & Ozbakkaloglu, T. (2020). A review of natural fiber composites: Properties, modification and processing techniques, characterization, applications. *Journal of Materials Science, 55*(3). https://doi.org/10.1007/s10853-019-03990-y

Guillebaud-Bonnafous, C., Vasconcellos, D., Touchard, F., & Chocinski-Arnault, L. (2012). Experimental and numerical investigation of the interface between epoxy matrix and hemp yarn. *Composites Part A: Applied Science and Manufacturing, 43*(11), 2046–2058. https://doi.org/10.1016/j.compositesa.2012.07.015

Hanipah, S. H., Omar, F. N., Talib, A. T., Mohammed, M. A. P., Baharuddin, A. S., & Wakisaka, M. (2020). Effect of silica bodies on oil palm fibre-polyethylene composites. *BioResources, 15*(1), 360–367. https://doi.org/10.15376/biores.15.1.360-367

Heydari-Meybodi, M., Saber-Samandari, S., & Sadighi, M. (2015). A new approach for prediction of elastic modulus of polymer/nanoclay composites by considering interfacial debonding: Experimental and numerical investigations. *Composites Science and Technology, 117*, 379–385. https://doi.org/10.1016/j.compscitech.2015.07.014

Hosseini, N., Javid, S., Amiri, A., Ulven, C., Webster, D. C., & Karami, G. (2015). Micromechanical viscoelastic analysis of flax fiber reinforced bio-based polyurethane composites. *Journal of Renewable Materials, 3*(3), 205–215. https://doi.org/10.7569/JRM.2015.634112

Hughes, M., Carpenter, J., & Hill, C. (2007). Deformation and fracture behaviour of flax fibre reinforced thermosetting polymer matrix composites. *Journal of Materials Science, 42*(7), 2499–2511. https://doi.org/10.1007/s10853-006-1027-2

Keck, S., & Fulland, M. (2016). Effect of fibre volume fraction and fibre direction on crack paths in flax fibre-reinforced composites. *Engineering Fracture Mechanics, 167*, 201–209. https://doi.org/10.1016/j.engfracmech.2016.03.037

Kern, W. T., Kim, W., Argento, A., Lee, E. C., & Mielewski, D. F. (2016). Finite element analysis and microscopy of natural fiber composites containing microcellular voids. *Materials and Design, 106*, 285–294. https://doi.org/10.1016/j.matdes.2016.05.094

Khaldi, M., Vivet, A., Bourmaud, A., Sereir, Z., & Kada, B. (2016). Damage analysis of composites reinforced with Alfa fibers: Viscoelastic behavior and debonding at the fiber/matrix interface. *Journal of Applied Polymer Science*, *133*(31). https://doi.org/10.1002/app.43760

Kiruthika, A. V. (2017). A review on physico-mechanical properties of bast fibre reinforced polymer composites. *Journal of Building Engineering*, *9*, 91–99. https://doi.org/10.1016/j.jobe.2016.12.003

Kong, C., Lee, H., & Park, H. (2016). Design and manufacturing of automobile hood using natural composite structure. *Composites Part B: Engineering*, *91*, 18–26. https://doi.org/10.1016/j.compositesb.2015.12.033

Mahboob, Z., & Bougherara, H. (2018). Fatigue of flax-epoxy and other plant fibre composites: Critical review and analysis. *Composites Part A: Applied Science and Manufacturing*, *190*, 440. https://doi.org/10.1016/j.compositesa.2018.03.034

Mahboob, Z., El Sawi, I., Zdero, R., Fawaz, Z., & Bougherara, H. (2017). Tensile and compressive damaged response in Flax fibre reinforced epoxy composites. *Composites Part A: Applied Science and Manufacturing*, *92*, 118–133. https://doi.org/10.1016/j.compositesa.2016.11.007

Miao, M., & Shan, M. (2011). Highly aligned flax/polypropylene nonwoven preforms for thermoplastic composites. *Composites Science and Technology*, *71*(15), 1713–1718. https://doi.org/10.1016/j.compscitech.2011.08.001

Mishra, V., & Biswas, S. (2013). Physical and mechanical properties of bi-directional jute fiber epoxy composites. *Procedia Engineering*, *51*, 561–566. https://doi.org/10.1016/j.proeng.2013.01.079

Modniks, J., & Andersons, J. (2010). Modeling elastic properties of short flax fiber-reinforced composites by orientation averaging. *Computational Materials Science*, *50*(2), 595–599. https://doi.org/10.1016/j.commatsci.2010.09.022

Modniks, J., & Andersons, J. (2013). Modeling the non-linear deformation of a short-flax-fiber-reinforced polymer composite by orientation averaging. *Composites Part B: Engineering*, *54*(1), 188–193. https://doi.org/10.1016/j.compositesb.2013.04.058

Mohammed, A. (2001). *Boundary Element Analysis–Theory and Programming*. London: CRC Press.

Mohammed, L., Ansari, M. N. M., Pua, G., Jawaid, M., & Islam, M. S. (2015). A review on natural fiber reinforced polymer composite and its applications. *International Journal of Polymer Science*, *2015*.

Nassar, M. M. A., Arunachalam, R., & Alzebdeh, K. I. (2017). Machinability of natural fiber reinforced composites: A review. *International Journal of Advanced Manufacturing Technology*, *88*(9–12), 2985–3004. https://doi.org/10.1007/s00170-016-9010-9

Naveen, J., Jawaid, M., Vasanthanathan, A., & Chandrasekar, M. (2018). Finite element analysis of natural fiber-reinforced polymer composites. In *Modelling of Damage Processes in Biocomposites, Fibre-Reinforced Composites and Hybrid Composites* (pp. 153–170). London: Elsevier. https://doi.org/10.1016/B978-0-08-102289-4.00009-6

Orifici, A. C., Herszberg, I., & Thomson, R. S. (2008). Review of methodologies for composite material modelling incorporating failure. *Composite Structures*, *86*(1–3), 194–210.

Panamoottil, S. M., Das, R., & Jayaraman, K. (2016). Anisotropic continuum damage model for prediction of failure in flax/polypropylene fabric composites. *Polymer Composites*, *37*(8), 2588–2597. https://doi.org/10.1002/pc.23453

Prajapati, H., Tevatia, A., & Dixit, A. (2021). Advances in natural fiber reinforced composites: A topical review. *Mechanics of Composite Materials*, *58*, 319–354.

Rafiquzzaman, M., Islam, M., Rahman, H., Talukdar, S., & Hasan, N. (2016). Mechanical property evaluation of glass—Jute fiber reinforced polymer composites. *Polymers for Advanced Technologies*, *27*(10), 1308–1316. https://doi.org/10.1002/pat.3798

Rahman, M. Z. (2017). Mechanical performance of natural fiber reinforced hybrid composite materials using finite element method based micromechanmincs and experiments. *Journal of Chemical Information and Modeling*, *53*(9), 1689–1699.

Ravandi, M., Teo, W. S., Tran, L. Q. N., Yong, M. S., & Tay, T. E. (2017). Low velocity impact performance of stitched flax/epoxy composite laminates. *Composites Part B: Engineering*, *117*, 89–100. https://doi.org/10.1016/j.compositesb.2017.02.003

Reddy, J. N. (2019). *Introduction to the Finite Element Method*. London: McGraw-Hill Education.

Reis, P. N. B., Ferreira, J. A. M., Antunes, F. V., & Costa, J. D. M. (2007). Flexural behaviour of hybrid laminated composites. *Composites Part A: Applied Science and Manufacturing*, *38*(6), 1612–1620. https://doi.org/10.1016/j.compositesa.2006.11.010

Saavedra Flores, E. I., & Haldar, S. (2016). Micro—Macro mechanical relations in Palmetto wood by numerical homogenisation. *Composite Structures*, *154*, 1–10. https://doi.org/10.1016/j.compstruct.2016.06.050

Scarponi, C. (2015). Hemp fiber composites for the design of a Naca cowling for ultra-light aviation. *Composites Part B: Engineering*, *81*, 53–63. https://doi.org/10.1016/j.compositesb.2015.06.001

Scarponi, C., & Messano, M. (2015). Comparative evaluation between E-Glass and hemp fiber composites application in rotorcraft interiors. *Composites Part B: Engineering*, *69*, 542–549. https://doi.org/10.1016/j.compositesb.2014.09.010

Sgriccia, N., Hawley, M. C., & Misra, M. (2008). Characterization of natural fiber surfaces and natural fiber composites. *Composites Part A: Applied Science and Manufacturing*, *39*(10), 1632–1637. https://doi.org/10.1016/j.compositesa.2008.07.007

Sharma, P., Priyanka, P., Mali, H. S., & Dixit, A. (2019). Geometric modeling and finite element analysis of kevlar monolithic and carbon-kevlar hybrid woven fabric unit cell. *Materials Today: Proceedings*, *26*, 766–774. https://doi.org/10.1016/j.matpr.2020.01.023

Shokrian, M. D., Shelesh-Nezhad, K., & Soudmand, B. H. (2016). 3D FE analysis of tensile behavior for co-PP/SGF composite by considering interfacial debonding using CZM. *Journal of Reinforced Plastics and Composites*, *35*(5), 365–374. https://doi.org/10.1177/0731684415622820

Singh, B., & Gupta, M. (2005). Natural fiber composites for building applications. *Natural Fibers, Biopolymers, and Biocomposites*, 261–290. https://doi.org/10.1201/9780203508206.ch8

Singh, D. K., Vaidya, A., Thomas, V., Theodore, M., Kore, S., & Vaidya, U. (2020). Finite element modeling of the fiber-matrix interface in polymer composites. *Journal of Composites Science*, *4*(2). https://doi.org/10.3390/jcs4020058

Sliseris, J., Yan, L., & Kasal, B. (2016). Numerical modelling of flax short fibre reinforced and flax fibre fabric reinforced polymer composites. *Composites Part B: Engineering*, *89*, 143–154. https://doi.org/10.1016/j.compositesb.2015.11.038

Sun, C. T., & Vaidya, R. S. (1996). Prediction of composite properties from a representative volume element. *Composites Science and Technology*, *56*(2), 171–179. https://doi.org/10.1016/0266-3538(95)00141-7

Sun, X., Gao, Z., Cao, P., Zhou, C., Ling, Y., Wang, X., Zhao, Y., & Diao, M. (2019). Fracture performance and numerical simulation of basalt fiber concrete using three-point bending test on notched beam. *Construction and Building Materials*, *225*, 788–800. https://doi.org/10.1016/j.conbuildmat.2019.07.244

Talreja, R. (1985). A continuum mechanics characterization of damage in composite materials. *Proceedings of the Royal Society of London. A. Mathematical and Physical Sciences*, *399*(1817), 195–216.

Távara, L., Moreno, L., Paloma, E., & Manti, V. (2019). Accurate modelling of instabilities caused by multi-site interface-crack onset and propagation in composites using the sequentially linear analysis and Abaqus. *Composite Structures*, *225*, 110993.

Tevatia, A., & Srivastava, S. K. (2015). Modified shear lag theory based fatigue crack growth life prediction model for short-fiber reinforced metal matrix composites. *International Journal of Fatigue*, *70*, 123–129. https://doi.org/10.1016/j.ijfatigue.2014.09.004

Wang, H., Lei, Y. P., Wang, J. S., Qin, Q. H., & Xiao, Y. (2016). Theoretical and computational modeling of clustering effect on effective thermal conductivity of cement composites filled with natural hemp fibers. *Journal of Composite Materials*, *50*(11), 1509–1521. https://doi.org/10.1177/0021998315594482

Wang, W., Lowe, A., Davey, S., Akhavan Zanjani, N., & Kalyanasundaram, S. (2015). Establishing a new forming limit curve for a flax fibre reinforced polypropylene composite through stretch forming experiments. *Composites Part A: Applied Science and Manufacturing*, *77*, 114–123. https://doi.org/10.1016/j.compositesa.2015.06.021

Wecławski, B. T., Fan, M., & Hui, D. (2014). Compressive behaviour of natural fibre composite. *Composites Part B: Engineering*, *67*, 183–191. https://doi.org/10.1016/j.compositesb.2014.07.014

Xiong, X., Hua, L., Miao, M., Shen, S. Z., Li, X., Wan, X., & Guo, W. (2018). Multi-scale constitutive modeling of natural fiber fabric reinforced composites. *Composites Part A: Applied Science and Manufacturing*, *115*, 383–396. https://doi.org/10.1016/j.compositesa.2018.10.016

Xiong, X., Shen, S. Z., Hua, L., Liu, J. Z., Li, X., Wan, X., & Miao, M. (2018). Finite element models of natural fibers and their composites: A review. *Journal of Reinforced Plastics and Composites*, *37*(9), 617–635. https://doi.org/10.1177/0731684418755552

Xue, D., Miao, M., & Hu, H. (2011). Permeability anisotropy of flax nonwoven mats in vacuum-assisted resin transfer molding. *Journal of the Textile Institute*, *102*(7), 612–620. https://doi.org/10.1080/00405000.2010.504566

Yan, L., Chouw, N., & Jayaraman, K. (2014). Flax fibre and its composites—A review. *Composites Part B: Engineering*, *56*, 296–317. https://doi.org/10.1016/j.compositesb.2013.08.014

Yang, L., Yan, Y., Liu, Y., & Ran, Z. (2012). Microscopic failure mechanisms of fiber-reinforced polymer composites under transverse tension and compression. *Composites Science and Technology*, *72*(15), 1818–1825. https://doi.org/10.1016/j.compscitech.2012.08.001

Yu, W. (2016). An introduction to micromechanics. *Applied Mechanics and Materials*, *828*, 3–24. https://doi.org/10.4028/www.scientific.net/amm.828.3

Zhang, B., Kawashita, L. F., & Hallett, S. R. (2020). Composites fatigue delamination prediction using double load envelopes and twin cohesive models. *Composites Part A: Applied Science and Manufacturing*, *129*, 105711.

Zhang, D., Milanovic, N. R., Zhang, Y., Su, F., & Miao, M. (2014). Effects of humidity conditions at fabrication on the interfacial shear strength of flax/unsaturated polyester composites. *Composites Part B: Engineering*, *60*, 186–192. https://doi.org/10.1016/j.compositesb.2013.12.031

Zhang, K. L., Zhang, J. Y., Hou, Z. L., Bi, S., & Zhao, Q. L. (2019). Multifunctional broadband microwave absorption of flexible graphene composites. *Carbon*, *141*, 608–617. https://doi.org/10.1016/j.carbon.2018.10.024

Zhi, C., Long, H., & Miao, M. (2017). Influence of microbond test parameters on interfacial shear strength of fiber reinforced polymer-matrix composites. *Composites Part A: Applied Science and Manufacturing*, *100*, 55–63. https://doi.org/10.1016/j.compositesa.2017.05.004

Zhong, Y., Tran, L. Q. N., Kureemun, U., & Lee, H. P. (2017). Prediction of the mechanical behavior of flax polypropylene composites based on multi-scale finite element analysis. *Journal of Materials Science*, *52*(9), 4957–4967. https://doi.org/10.1007/s10853-016-0733-7

Zou, Z., Reid, S. R., & Li, S. (2003). A continuum damage model for delaminations in laminated composites. *Journal of the Mechanics and Physics of Solids*, *51*(2), 333–356. https://doi.org/10.1016/S0022-5096(02)00075-3

9 Lead-Free Multiferroic BiFeO₃ Based Sustainable Green Composites
Applications, Opportunities and Future Challenges

Manish Kumar, Arvind Kumar, Satyam Kumar, and Z. R. Khan

CONTENTS

9.1 INTRODUCTION

Our need of technological development in turn leads to the urge to improve upon existing and obsolete novel materials, and often results in new materials and innovation. The multiferroic family is one such eminent class of materials. Multiferroism in the materials justifies with the simultaneous existence of more than one ferroic orders such as ferroelectricity, magnetism, ferroelasticity and ferrotoroidicity. Studies of their synthesis, structure and physical characterization have been the subject of widespread interest during the recent past. The growing interest in this field is not due mainly to the academic curiosity but also due to the fact that it has been realized that much of the rapidly increasing demand of industry and technology can be met with, in a thorough and elegant fashion by the exploitation of their unique physical properties like

DOI: 10.1201/9781003272625-9

ferroelectricity, ferromagnetism and superconductivity. Microminiaturized devices, memory devices, magnetic media (discs and tapes) for data storage, magnetic field sensors etc. are a few examples where these materials are finding potential application due to their excellent electrical and magnetic properties (Spaldin & Ramesh, 2019; Chandra, 2019). Out of a number of multiferroic materials, $BiFeO_3$ (BFO) is a perovskite type multiferroic material having the ferroelectricity as well as antiferromagnetic nature near room temperature (Matsukura & Ohno, 2015; Fiebig et al., 2016). It has distorted rhombohedral structure due to which small but countable magnetic property exists in it simultaneously with the ferroelectric nature. The magnetic and ferroelectric transition temperatures of BFO are reported 1103K and 643K, respectively (Tokura & Nagaosa, 2017). There are a number of reports available about its usefulness due to its excellent multiferroic nature in bulk and nano forms (Ramesh & Spaldin, 2007; Khomskii, 2006; Hill, 2000; Khomskii, 2001; Gupta et al., 2015). In single phase as well as in composite phases with various ferroelectric materials, such as $BaTiO_3$, $PbTiO_3$ etc., it shows magnetoelectric coupling i.e., control on magnetization via applied electric field and/control on polarization via applied magnetic field. In the extension of this magnetoelectric coupling which is extremely useful in spintronic device applications, people have reported magnetodielectric coupling in BFO-based materials and confirmed the evidence of magnetoelectric coupling via the independent dielectric and magnetic measurements (Kumar et al., 2013, 2014; Shankar et al., 2014). They have obtained the dielectric transition via temperature dependent dielectric measurement and magnetic transition via magnetic measurement and check the closeness of both the transitions. If the magnetic transition is close to dielectric anomaly in the dielectric vs. temperature plots, then it is decided that there are some evidences of magnetoelectric coupling in the material. Apart from the spintronic applications, the BFO-based materials are extremely useful in the photovoltaic applications as absorber layer because of the excellent optical properties such as bandgap of the BFO in the range which is required for the photovoltaic applications (Zhang et al., 2019; Nechache et al., 2015).

In the present chapter, efforts have been made to present the BFO-based composites with their synthesis as well as multiferroic properties analysis. BFO-based lead-free green composites are very useful in the various device applications, so, we are also committed to present their usefulness in the device applications such as photovoltaic applications especially in the perovskite solar cells. Efforts also have been made to discuss the use of BFO-based materials as absorber layer in the perovskite solar cells heterostructures including the new opportunities and future challenges.

9.2 SYNTHESIS TECHNIQUES FOR COMPOSITE FORMATION

Synthesis is a very crucial part of the materials research. After successful synthesis of material composites were fabricated and then tested for mechanical performance. But a number of factors are important during the synthesis such as control on shape and size during the synthesis of nanomaterials, mono disparity, surface morphology and fictionalization etc. There are usually various methods to synthesize the BFO-based materials such as Sol-gel, Co-precipitation and solid state reaction etc. (Sahni et al., 2021; Muneeswaran et al., 2013; Sharma et al., 2020). A lot of literature is available for

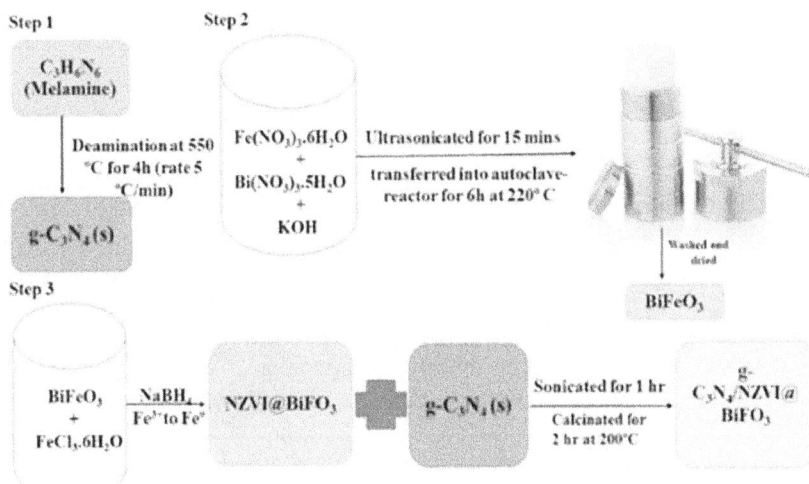

FIGURE 9.1   A schematic depiction for the synthesis of BiFeO$_3$ and NZVI@BiFeO$_3$/g-C$_3$N$_4$ composite.

Source: Reproduced with permission from Rahman et al. (2021)

the composite formation via the various techniques such as Mohrana et al. who synthesized hydroxylated bismuth ferrite-Poly (vinylidene fluoride-co-hexafluoropropylene) composites via conventional solid state reaction method for energy storage devices (Moharana et al., 2016). They have Bi_2O_3 and Fe_2O_3 in a proper stoichiometric ration and thoroughly mixed by an agate mortar and pestle followed by calcination. After that BFO-PVDF-HFP composite film was prepared via solution casting technique. Rahman et al. have synthesized novel graphitic carbon nitride based nano zero-valent iron doped bismuth ferrite ternary composite [photo-catalysts (g-C3N4 and BiFeO$_3$) and composite photo-catalyst (NZVI@BiFeO$_3$/g-C$_3$N$_4$)]. Finally, they have prepared composites of NZVI@BiFeO$_3$ and g- C3N4 with the mixing of distilled water followed by drying as well as calcination as shown in Figure 9.1 (Rahman et al., 2021).

9.3   CHARACTERIZATION TECHNIQUES

9.3.1   STRUCTURAL ANALYSIS

BFO nanostructures and composites have demonstrated various strategic characteristics such as optical, electrical, magnetic and valley properties and the strong correlation among them. In addition, bulk BFO showed noticeable photovoltaic effect soon after BFO-based hetrostructures have been under extensive investigation. BFO bulk material has possessed rhombohedral symmetry with space group-R3c and lattice constant a = 5.63 Å and the rhombohedral angel 89.45° at room temperature (Singh et al., 2014; Qi et al., 2017). So far, numerous research groups in different forms have reported BFO structural properties viz. ceramics, nanostructures and thin films. Mohanty et al. have synthesized BFO powder and investigated purity and phase of synthesized materials using X-ray diffraction patterns. The structural parameters of grown materials were analyzed using the Rietveld refinement with the help of MAUD software, which confirms the orthorhombic phase with *Pbnm* space group. The calculated lattice parameters a = 5.6347 Å, b = 5.6071 Å and c = 7.9175 Å that are well agreed with standard data card: 01–084–1089. The average crystallites size (D) of ceramic powder has been calculated corresponding to the intense peak and obtained in range of 1.17 μm (Mohanty et al., 2021). Najm et al. have grown 'In' substituted BFO ceramics via solid state reactions method. The phase and impurity of grown ceramics has been investigated through X-ray diffraction patterns. XRD peaks revealed fine crystallization along preferred orientation 021, 104, 110, 202 and 024 and close agreement with rhombohedral crystal structural with space group R3c 161, standard data of diffraction ICSD NO: 980066143. They were observed BFO phase reduction as incorporation of 'In' in BFO lattices. The lattice parameters of BFO show the reduction at x = 0.1 and further increasing the doping concentrations x = 0.2–0.6 lattice parameters becomes enhanced. This may be due to the In^{+3} has substituted with Fe^{+2} as a result increment in lattice parameters (Najm et al., 2021). Similarly, thin films structural properties of BFO also have been studied in detail. X. Deng et al. have studied structural properties of Ni doped BFO thin films and found polycrystalline peaks of the rhombohedral phase (JCPDS-72–2422), with space group R3c_161 (Deng et al., 2020). Recently, our group (Chaudhary et al.) prepared composite of $CoFe_2O_4$ with a fixed composition of $BiFeO_3$-$BaTiO_3$ solid solution via conventional solid state reaction method and confirmed the composite phase formation of the multiferroic-magnetic lead-free composite (Chaudhary et al., 2019). Kumar et al. (2018) have also reported BFO-BT composites via solid state reaction route and confirm the magnetic property enhancement in the composites. They have confirmed the impurity free phase in the BFO-BT composites via powder XRD pattern with different CuO and MnO_2 additives as shown in Figure 9.2.

In order to understand the lattice vibrational mode of BFO, Raman spectroscopy was employed because it is a non-destructive chemical analysis technique which provides comprehensive information about the phase, polymorph, crystalinity and molecular interactions. Raman spectroscopy is a light scattering based phenomenon from an intense laser light. X. Deng et al. have analyzed the effect of Ni doping on the phase of BFO. They found rhombohedrally distorted BFO with R3c space group exhibits 13 ($\Gamma = 4A_1 + 9E$) Raman active mode. The eight clear modes were observed in all compositions of films. It is well known that low frequency modes relate to the Bi-O vibration and the high frequency E mode relates to the Fe-O vibration (Deng et al., 2020).

Surface morphological properties of BFO nanostructures, thin films and nano composites have been investigated via scanning electron microscope. Recently, high-resolution scanning electron microscopes have been available to characterize and investigate the surface morphological properties of BFO materials. K. P. Remya et al. have investigated the impact of morphology on physical and chemical properties of perovskite nanostructures. The variation of morphology can be seen from SEM images with different growth conditions of BFO nanostructures nanorods, nanoparticle, nanotube etc. (Remya et al., 2020). M. Sakar et al. have investigated hierarchical bismuth ferrite ($BiFeO_3$/BFO) nanostructures through the FESEM. They obtained an average size of particulate around 400 nm and the average thickness of the nano flakes is about 60 nm (Sakar & Balakumar, 2017).

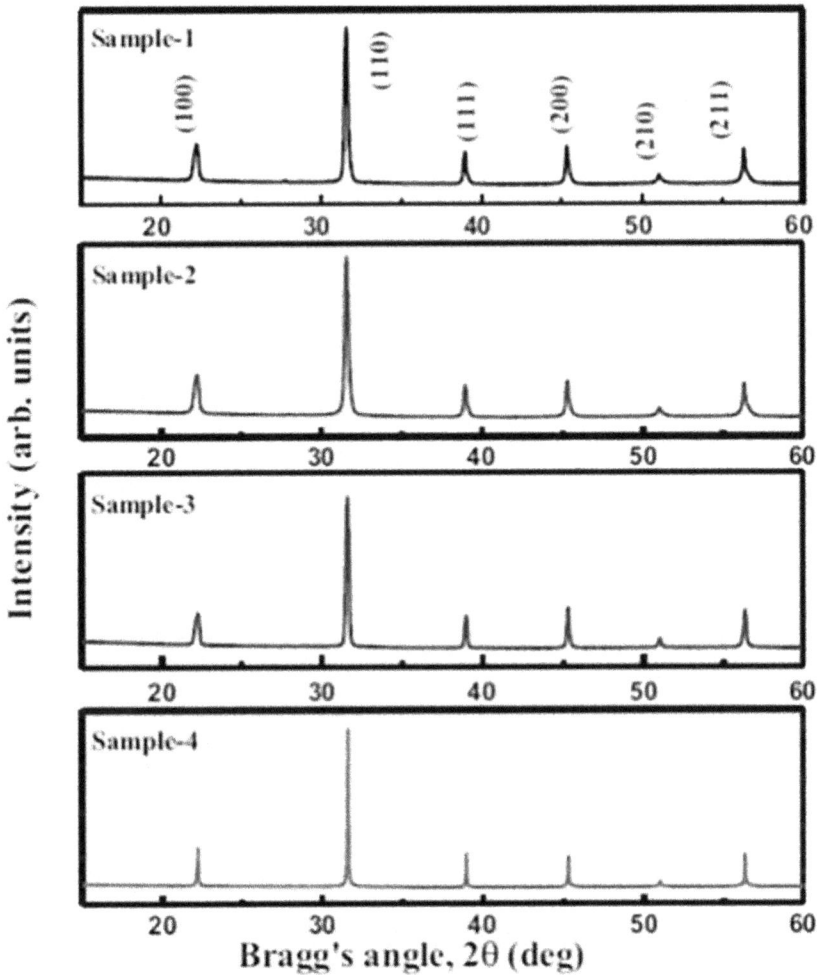

FIGURE 9.2   Powder XRD patterns of $0.675BiFeO_3$–$0.325BaTiO_3$ composite prepared via CuO and MnO_2 additives.

Source: Reproduced with permission from Kumar et al. (2018)

9.3.2   MULTIFERROIC PROPERTIES ANALYSIS

The direct multiferroic properties can be confirmed from the simultaneous presence of ferroelectric as well as magnetic phases via P-E loop and M-H loop measurements in the materials near room temperature. In the BFO-BT composites, Kumar et al. have reported the enhancement and simultaneous phase presence of ferroelectric as well as magnetic orderings as shown in the Figure 9.3 (Kumar et al., 2018). They have reported the enhancement in the multiferroic properties of BFO-BT composites, namely 1–4 [$0.675BiFeO_3$–$0.325BaTiO_3$ and added x wt%CuO + y wt%MnO₂ varied

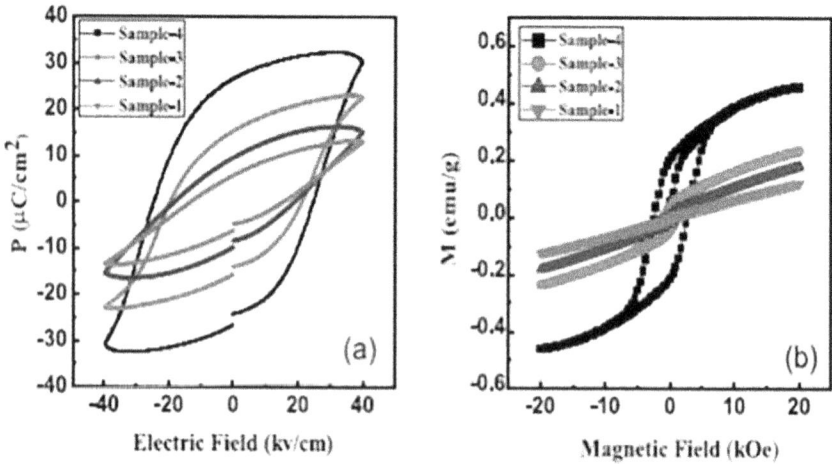

FIGURE 9.3   (a) Ferroelectric P *vs.* E hysteresis loops (samples 1–4) and (b) M *vs.* H hysteresis loops (samples 1–4).

Source: Reproduced with permission from Kumar et al. (2018)

as (1) $x = 0$, $y = 0$; (2) $x = 0.1$, $y = 0.15$; (3) $x = 0.2$, $y = 0.30$; and (4) $x = 0.4$, $y = 0.6$.] with the addition of additives CuO and MnO_2 and confirmed the correspondence in between enhanced grain size and domain switching. Along with the ferroelectric nature (Figure 9.3(a)), magnetic nature is also enhanced with the addition of additive in the host composite as shown in Figure 9.3(b) (Kumar et al., 2018).

9.3.3 MAGNETOELECTRIC COUPLING ANALYSIS

Magnetoelectric coupling is the coupling between ferroelectric and magnetic orderings of materials. It has a pronounced interest as per the applications point of view in the spintronics, photovoltaics devices etc. In this coupling, the magnetization is easily controlled without magnetic field i.e., with only electric field or vice versa, which offers the possibilities for novel magnetoelectric coupling based memory and spintronic devices. The BFO-based BFO-BT lead-free composite is an interesting composite material and important to study for the magnetoelectric coupling point of view. Independently, the BFO has low magnetic character and shows the antiferromagnetic character. So, in the light of it, BT is added in this material and efforts are looking for better magnetic properties in the multilayer structures. It is believed that the magnetic interaction between BFO and BT layers at the interface may enhance the magnetic moment in comparison to individual single phase layers. Recently, Lazenka et al. have studied the effect of strain and interfaces on the magnetoelectric coupling of BFO-BT multilayer structures and reported the magnetoelectric coupling in BFO-BT multilayers reaches the magnetoelectric coupling coefficient value ~ 20.75 V/cm.Oe [31]. They have used the direct longitudinal AC method for the calculation of the magnetoelectric coupling coefficient value with respect to the static magnetic field. It is

(a)

(b)

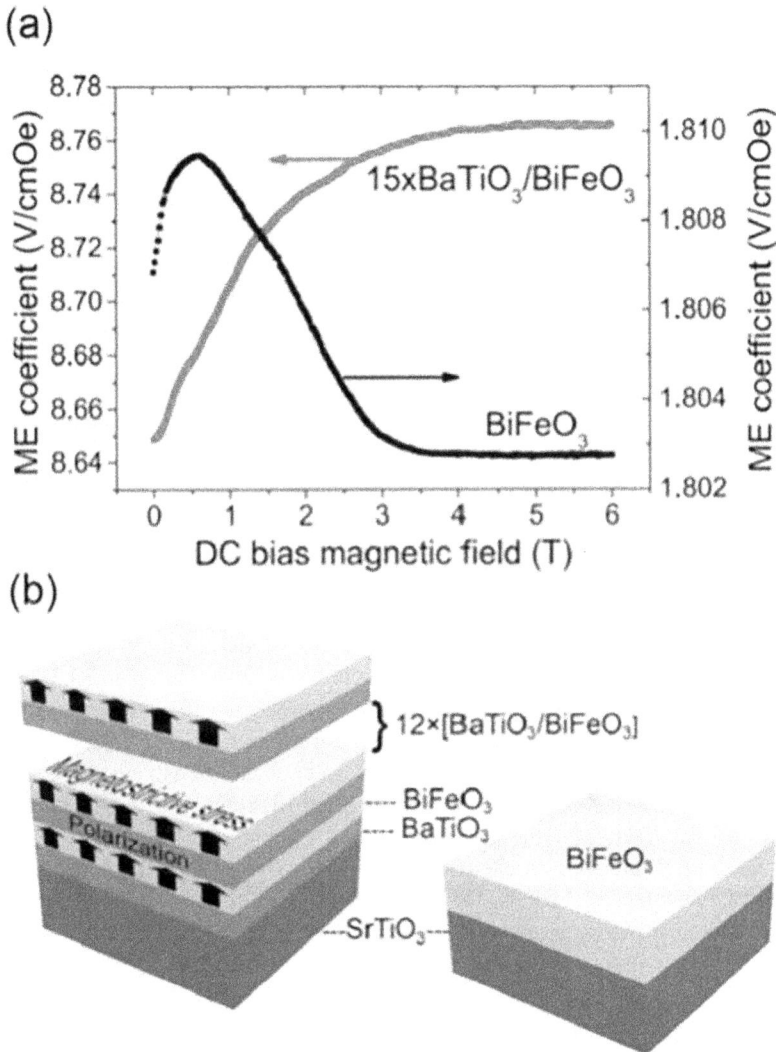

FIGURE 9.4 (a) Magnetoelectric coupling coefficient *vs.* DC magnetic field near room temperature for BFO-BT multilayer and the BFO thin film. (b) Schematic multilayer structure of a ferroelectric-multiferroic (left) and $BiFeO_3$ thin film (right).

Source: Reproduced with permission from Lazenka et al. (2015)

confirmed from Figure 9.4(a) that the magnetoelectric coupling coefficient for multilayer BFO-BT structure is larger than the single phase BFO thin film. Figure 9.4(b) shows the schematic representation of the multilayer (left) and BFO thin film (right) structures representing the improvement in the magnetoelectric coupling coefficient via magnetostrictive-piezoelectric interface coupling (Lazenka et al., 2015).

9.4   APPLICATIONS OF BFO IN PHOTOVOLTAICS

In view of the increasing global energy crisis and environmental issues across the world, the development of clean and sustainable energy sources is required to fulfill energy needs for present day applications. In various available energy sources, solar energy has been found to be one of the most important and valuable renewable energy sources due to its clean and pollution free approach (Correa-Baena et al., 2017; Yang & Alexe, 2018). In the recent past various series of materials have been studied theoretically and experimentally to predict the applicability of materials in photovoltaic cell with efficient and maximum power conversion efficiency. Presently, in the solar cell market, highly used and stable commercialized crystalline silicon solar panels occupy the maximum shares with maximum power conversion efficiency (> 26%) (Correa-Baena et al., 2017; Green et al., 2018; Lopez-Varo et al., 2016). But the processing and manufacturing cost is high. In a series of developments, second and third generation semiconductor solar cells, thin film structures, amorphous silicon solar cells, dye-sensitized solar cells, quantum dot solar cells and organic-inorganic hybrid perovskite solar cells have been recently explored for the cost-effective photovoltaic cell applications (Chen et al., 2017; Ansari et al., 2018; Bhatt & Lee, 2017; Hou et al., 2018; Yang & Jen, 2017). But typical p-n junction diode-based semiconductor solar cells have many disadvantages such as limitation on V_{oc}, small short circuit current and limitation in their power conversion efficiency (PCE) due to the Shockley-Queisser limit. Such problems limit the typical p-n junction semiconductor diodes to convert more than 33.7% of the incident light (Tan et al., 2016; Ren et al., 2017; Paillard et al., 2016). As an alternative approach to remove all the drawbacks of typical p-n junction diodes, ferroelectric (FE) photovoltaic effects gained much attention (Yang et al., 2012; Huang, 2010; Zhang et al., 2017; Jalaja & Dutta, 2017; Bhatnagar et al., 2013; Yuan et al., 2014). This effect provides a great potential in the development of the photovoltaic devices with higher PCE value with its unique characteristics. Apart from it, it was noted that ferroelectric materials are inexpensive, abundant and stable. These qualities favor making ferroelectric photovoltaic devices for the commercial application.

Ferroelectric, $BiFeO_3$ based nanomaterial, thin films and heterostructures have gained much interest now days due to their potential applications and physical properties among the available class of perovskite materials. In the recent past several synthesis methods and characterization techniques were explored on BFO-based structures for the detailed study of their physical properties and possible applications in the present-day technology. $BiFeO_3$ based structures were found to be a better alternative of lead-free perovskite structures, and display room-temperature multiferroic properties and proper band gaps etc. (Biswas et al., 2018; Bibes & Barthélémy, 2008; Wu et al., 2016). These physical properties make it applicable for spintronic, electronic and opto-electronic devices etc. BFO material has significant response for its applications in both of its physical forms (nano or bulk) due to its proper band gap (< 2.7 eV), high temperature ferroelectric and magnetic ordering (Biswas et al., 2018; Bibes & Barthélémy, 2008; Wu et al., 2016). Such interesting properties of the BFO among the various other combinations of the perovskites make its a suitable candidate for developing ferroelectric photovoltaic devices with high power conversion efficiency. In the past decades various studies were performed on BFO for its application in photovoltaic

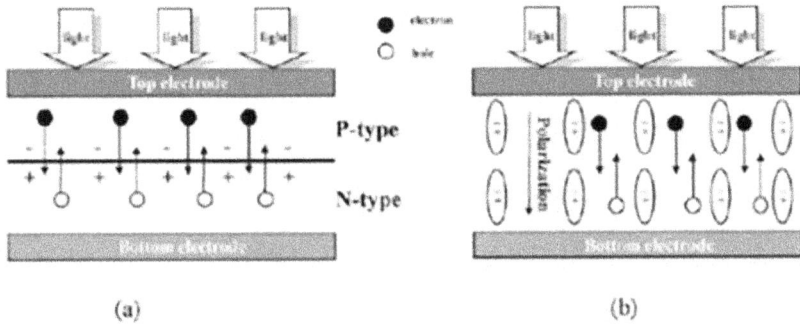

FIGURE 9.5   Schematic illustration of (a) p-n junction photovoltaic and (b) ferroelectric photovoltaic cell.

Source: Reproduced with permission from Chen et al. (2019)

devices. Typically, photovoltaic effect comes from the two basic processes, i.e., excitation of electron–hole (e–h) pairs and separating the e–h pairs. Figure 9.5 shows the illustration of the working principle of traditional p-n junction photovoltaic cell and ferroelectric photovoltaic cell, respectively (Chen et al., 2019). BFO is demonstrated for photovoltaic effect in several ways as discussed later:

9.4.1   PHOTOVOLTAIC EFFECTS IN BFO-BASED BULK AND NANOMATERIAL

Choi et al. in 2009 explored and demonstrated the diode and photovoltaic effects in single ferroelectric-domain bulk BFO (Choi et al., 2009). Here, light-induced e–h pairs are separated by the depolarized field of ferroelectric BFO. Figure 9.6 (a–c) represented the schematic layered structure of BFO along with the measured J-V curves for the applications in photovoltaic devices. It was observed that due to the small optical band gap (~2.2–2.8 eV) of BFO (Gao et al., 2006; Himcinschi et al., 2015; Sharma et al., 2015), its photovoltaic effect lies in the visible light region of the em radiation which is required and important for its practical applications in the photovoltaic devices. Large value of the V$_{oc}$ is observed due to depolarized field of BFO in photovoltaic metal/BFO/metal structures. BFO-based heterostructures were further studied to see the effect of optical modulation on the magneto electric properties.

In 2012 Hung et al. investigated the photovoltaic characteristics in bulk BiFeO$_3$ based photovoltaic devices (Hung et al., 2012). They found that the photovoltaic response is dependent on different factors such as wavelength of light, light intensity and sample thickness etc. It was also found that samples with lower thickness display better device parameters regarding its application point of view, like large value of open circuit voltage (V$_{oc}$) and short circuit current density etc. They reported stronger photovoltaic response at wavelength, $\lambda = 373$ nm and at $\lambda = 532$ nm. Further, to increase the photovoltaic response of the bulk BFO-based devices, Tu et al. and other researchers used doped BFO by different metal ions at A sites (Tu et al., 2013, 2015; Hung et al., 2014). In preceding years BFO-based nanomaterials in different structural

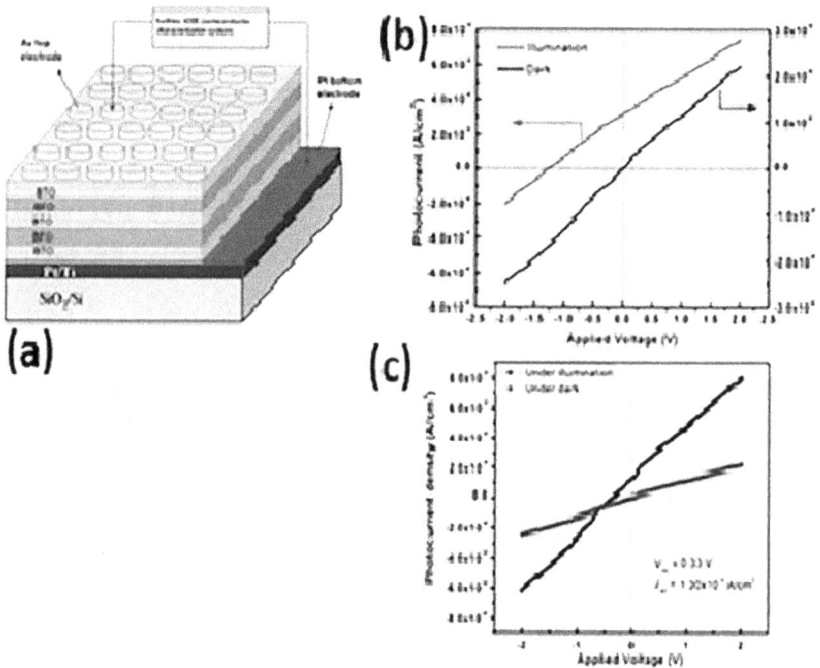

FIGURE 9.6   (a) BFO-BT multilayer as metal/ferroelectric/metal capacitor depiction, (b) BFO-BT multilayer J-V curves, (c) single layer BFO J-V curve (under dark and illumination, $\lambda = 405$ nm with a intensity 160 mW cm^{-2}).

Source: Reproduced with permission from Sharma et al. (2015)

shapes (nanoparticles, nanotubes, nanofibers and nanowires etc.) were explored for possibility with higher efficiency photovoltaic applications. Sudakar et al. (Mocherla et al., 2013) tuned the band gap of BFO from 2.32 eV to 2.09 eV nanoparticles by controlling the size variation and demonstrated the photovoltaic parameters. They found better PCE value and parameters for nanosized BFO in devices as compared to the bulk BFO in the photovoltaic devices. Figure 9.7 (a–c) shows the measurement set up of random BFO nanofiber-based photovoltaic devices and corresponding current-voltage characteristics as demonstrated by (Fei et al., 2015). They found nanofibers showing enhanced photo-voltaic properties as compared to thin film structures.

9.4.2   PHOTOVOLTAIC EFFECTS IN BFO-BASED THIN FILM STRUCTURES

Apart from the bulk and nanosized BFO-based photovoltaic devices, to further improve the photovoltaic performance of BFO-based materials, different research groups prepared various BFO and doped BFO thin films based devices. It was observed that PV effect in BFO-based films was majorly dominated by photo-excited electrical carrier across junction due to the much larger short-circuit photo-current under green light than red light illumination. It is also sensitive to the ferroelectric domain wall configuration,

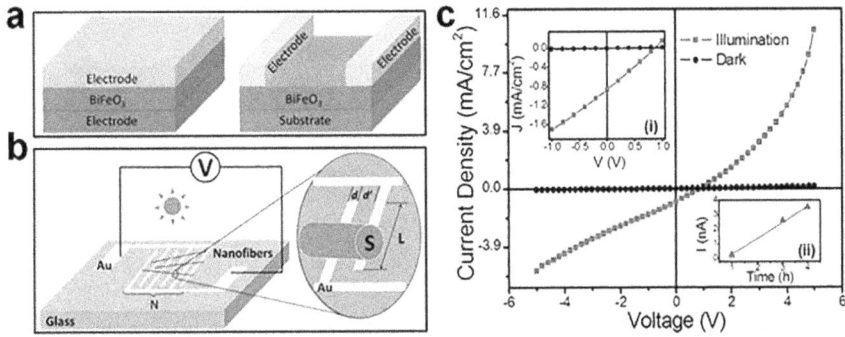

FIGURE 9.7   (a) Thin film-based photovoltaic devices schematic diagram (left device is with a parallel capacitor configuration whereas right one shows laterally aligned inter-digital electrodes). (b) Random BFO nano fiber-based photovoltaic devices schematic diagram and (c) nanofibers structure J-V curve. (In the inset, expanded outlook of J-V curves nearby zero-bias and averaged photo-current after numerous measurements for different deposition time (1 to 4 h).)

Source: Reproduced with permission from Fei et al. (2015)

oxygen vacancy concentration and electrodes etc. Doping in BFO material serves the purpose of reducing band gap to enhance the optical absorption response in the visible part of the light spectrum. Gao et al. investigate the photovoltaic response of La doped BFO films grown on $La_{0.7}Sr_{0.3}MnO_3/SrTiO_3$ substrates and Ag as top electrode (Gao et al., 2014a, 2014b). They found good open circuit voltage with and without illumination of light. Further, Cao et al. studied in detail the performance of the device (Gao et al., 2016). Figure 9.8 shows the BFO films on commercial Pt(1 1 1)/Ti/SiO₂/Si(1 0 0) substrates and studied in detail with variation of BFO thickness (t), growth temperature (Tg) for PV properties under various illumination intensity (Chang et al., 2013). It was seen from the investigation that layered structure of BFO films shows significant photovoltaic phenomena with thickness and growth temperature variation.

9.5   CHALLENGES AND FUTURE ASPECTS OF BFO-BASED PHOTOVOLTAIC DEVICES

It has been observed that in the last few decades various BFO-based structures were studied for the commercial application of the photovoltaic devices. However, the power conversion efficiency (PCE) of many BFO-based photovoltaic devices is still too low. The possible reasons for this may be due to its large band gap value as compared to the ideally predicted value of ~ 1.4 eV for higher PCE value. Another possible reason for low PCE value may be due to its insulating behaviour causing small value of output photocurrent. Thus, based on the earlier description, it is clear that to improve the PCE more attention should be devoted towards the BFO-based photovoltaic devices. It is required to develop highly crystalline BFO films with minimum defects, good range of band gap by doping of suitable elements which could allow more absorption of light

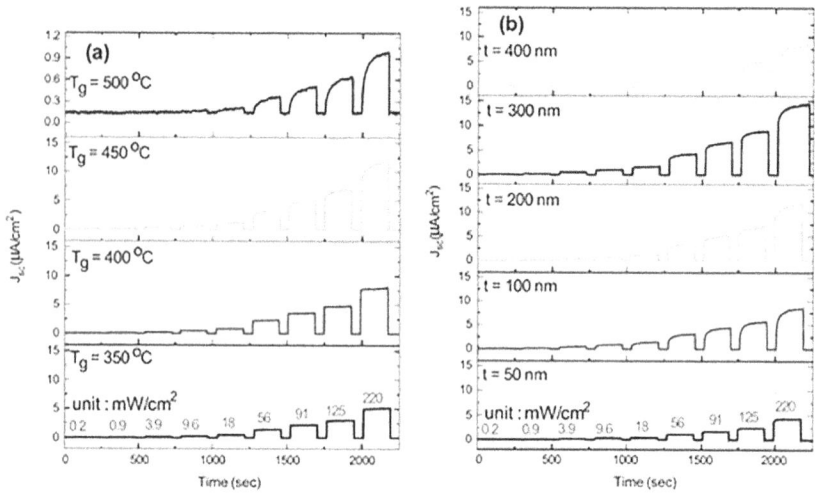

FIGURE 9.8   Photocurrent (time-dependent) under laser illumination of k = 405 nm at numerous intensity values for (a) BFO films deposited on Pt/Ti/SiO$_2$/Si(1 0 0) substrates (200-nm-thick) at numerous T$_g$ and (b) BFO films grown on Pt/Ti/SiO$_2$/Si(1 0 0) substrates at Tg = 450 °C with dissimilar BFO thickness.

Source: Reproduced with permission from Chang et al. (2013)

in visible and ultraviolet regions etc. Overall, more dedication is required to fabricate stable BFO-based ferroelectric photovoltaic devices with higher PCE along with better photovoltaic parameters (such as large open-circuit voltage, high short-circuit etc.).

9.6   CONCLUDING REMARKS

In summary, we have discussed the lead-free green BFO-based composites for the applications point of view. BFO-based materials are very useful in the spintronic memory based applications as well as photovoltaics because of the simultaneous presence of two or more ferroic orders and control on magnetization by applied electric field and/or vice-versa. Apart from that we have addressed some of the fabrication and characterization techniques of lead-free multiferroic bismuth ferrite based green composites. Efforts also have been made to discuss the use of BFO-based materials in the photovoltaic applications in detail on the basis of photovoltaic effects in BFO-based bulk, nanomaterial, thin film structures including the new challenges and future aspects of BFO-based photovoltaic devices.

REFERENCES

Ansari, M. I. H., Qurashi, A., & Nazeeruddin, M. K. (2018). Frontiers, opportunities, and challenges in perovskite solar cells: A critical review. *Journal of Photochemistry and Photobiology C: Photochemistry Reviews, 35,* 1–24.

Bhatnagar, A., Chaudhuri, A. R., Kim, Y. H., Hesse, D., & Alexe, M. (2013). Role of domain walls in the abnormal photovoltaic effect in BiFeO$_3$. *Nature Communications, 4*(1), 1–8.

Bhatt, M. D., & Lee, J. S. (2017). Current progress and scientific challenges in the advancement of organic—Inorganic lead halide perovskite solar cells. *New Journal of Chemistry, 41*(19), 10508–10527.

Bibes, M., & Barthélémy, A. (2008). Towards a magnetoelectric memory. *Nature Materials, 7*(6), 425–426.

Biswas, P. P., Thirmal, C., Pal, S., & Murugavel, P. (2018). Dipole pinning effect on photovoltaic characteristics of ferroelectric BiFeO$_3$ films. *Journal of Applied Physics, 123*(2), 024101.

Chandra, P. (2019). Multifunctionality goes quantum critical. *Nature Materials, 18*(3), 197–198.

Chang, H. W., Yuan, F. T., Yu, Y. C., Chen, P. C., Wang, C. R., Tu, C. S., & Jen, S. U. (2013). Photovoltaic property of sputtered BiFeO$_3$ thin films. *Journal of Alloys and Compounds, 574*, 402–406.

Chaudhary, P., Kumar, M., Dabas, S., & Thakur, O. P. (2019). Enhanced magneto-electric coupling and energy storage analysis in (BiFeO$_3$—BaTiO$_3$)/CoFe$_2$O$_4$ composites. *Journal of Materials Science: Materials in Electronics, 30*(15), 13910–13923.

Chen, B., Zheng, X., Bai, Y., Padture, N. P., & Huang, J. (2017). Progress in tandem solar cells based on hybrid organic—Inorganic perovskites. *Advanced Energy Materials, 7*(14), 1602400.

Chen, G., Chen, J., Pei, W., Lu, Y., Zhang, Q., Zhang, Q., & He, Y. (2019). Bismuth ferrite materials for solar cells: current status and prospects. *Materials Research Bulletin, 110*, 39–49.

Choi, T., Lee, S., Choi, Y. J., Kiryukhin, V., & Cheong, S. W. (2009). Switchable ferroelectric diode and photovoltaic effect in BiFeO$_3$. *Science, 324*(5923), 63–66.

Correa-Baena, J. P., Saliba, M., Buonassisi, T., Grätzel, M., Abate, A., Tress, W., & Hagfeldt, A. (2017). Promises and challenges of perovskite solar cells. *Science, 358*(6364), 739–744.

Deng, X., Zeng, Z., Gao, R., Wang, Z., Chen, G., Cai, W., & Fu, C. (2020). Study of structural, optical and enhanced multiferroic properties of Ni doped BFO thin films synthesized by sol-gel method. *Journal of Alloys and Compounds, 831*, 154857.

Fiebig, M., Lottermoser, T., Meier, D., & Trassin, M. (2016). The evolution of multiferroics. *Nature Reviews Materials, 1*(8), 1–14.

Fei, L., Hu, Y., Li, X., et al. (2015). Electrospun bismuth ferrite nanofibers for potential applications in ferroelectric photovoltaic devices. *ACS Applied Materials & Interfaces, 7*(6), 3665–3670.

Gao, R. L., Fu, C., Cai, W., Chen, G., Deng, X., & Cao, X. (2016). Switchable photovoltaic effect in Au/Bi0.9La0.1FeO3/La0.7Sr0.3MnO3 heterostructures. *Materials Chemistry and Physics, 181*, 277–283.

Gao, R. L., Yang, H. W., Chen, Y. S., Sun, J. R., Zhao, Y. G., & Shen, B. G. (2014a). Oxygen vacancies induced switchable and non-switchable photovoltaic effects in Ag/Bi0.9La0.1FeO3/La0.7Sr0.3MnO3 sandwiched capacitors. *Applied Physics Letters, 104*(3), 031906.

Gao, R. L., Yang, H. W., Chen, Y. S., Sun, J. R., Zhao, Y. G., & Shen, B. G. (2014b). The study of open circuit voltage in Ag/Bi0.9La0.1FeO3/La0.7Sr0.3MnO3 heterojunction structure. *Physica B: Condensed Matter, 432*, 111–115.

Gao, F., Yuan, Y., Wang, K. F., Chen, X. Y., Chen, F., & Liu, J.-M. (2006). Preparation and photoabsorption characterization of BiFeO$_3$. *Applied Physics Letters, 89*, 102506.

Green, M. A., Hishikawa, Y., Dunlop, E. D., Levi, D. H., Hohl-Ebinger, J., & Ho-Baillie, A. W. (2018). Solar cell efficiency tables (version 52). *Progress in Photovoltaics: Research and Applications, 26*(7), 427–436.

Gupta, R., Chaudhary, S., & Kotnala, R. K. (2015). Interfacial charge induced magnetoelectric coupling at BiFeO$_3$/BaTiO3 bilayer interface. *ACS Applied Materials & Interfaces, 7*(16), 8472–8479.

Hill, N. A. (2000). Why are there so few magnetic ferroelectrics? *The Journal of Physical Chemistry B, 104*(29), 6694–6709.

Himcinschi, C., Bhatnagar, A., Talkenberger, A., et al. (2015). Optical properties of epitaxial BiFeO$_3$ thin films grown on LaAlO3. *Applied Physics Letters, 106*(1), 012908.

Hou, J., Inganäs, O., Friend, R. H., & Gao, F. (2018). Organic solar cells based on non-fullerene acceptors. *Nature Materials, 17*(2), 119–128.

Huang, H. (2010). Ferroelectric photovoltaics. *Nature Photonics, 4*(3), 134–135.

Hung, C. M., Tu, C. S., Xu, Z. R., Chang, L. Y., Schmidt, V. H., Chien, R. R., & Chang, W. C. (2014). Effect of diamagnetic barium substitution on magnetic and photovoltaic properties in multiferroic BiFeO$_3$. *Journal of Applied Physics, 115*(17), 17D901.

Hung, C. M., Tu, C. S., Yen, W. D., Jou, L. S., Jiang, M. D., & Schmidt, V. H. (2012). Photovoltaic phenomena in BiFeO$_3$ multiferroic ceramics. *Journal of Applied Physics, 111*(7), 07D912.

Jalaja, M. A., & Dutta, S. (2017). Switchable photovoltaic properties of multiferroic KBiFe2O5. *Materials Research Bulletin, 88*, 9–13.

Khomskii, D. I. (2001, March). Magnetism and ferroelectricity; why do they so seldom coexist? In *APS March Meeting Abstracts* (pp. C21–002). Washington State Convention Center Seattle, Washington: Harvard University Press.

Khomskii, D. I. (2006). Multiferroics: Different ways to combine magnetism and ferroelectricity. *Journal of Magnetism and Magnetic Materials, 306*(1), 1–8.

Kumar, A., Narayan, B., Pachat, R., & Ranjan, R. (2018). Magnetic enhancement of ferroelectric polarization in a self-grown ferroelectric-ferromagnetic composite. *Physical Review B, 97*(6), 064103.

Kumar, M., Shankar, S., Kotnala, R. K., & Parkash, O. (2013). Evidences of magneto-electric coupling in BFO—BT solid solutions. *Journal of Alloys and Compounds, 577*, 222–227.

Kumar, M., Shankar, S., Parkash, O., & Thakur, O. P. (2014). Dielectric and multiferroic properties of 0.75 BiFeO$_3$–0.25 BaTiO$_3$ solid solution. *Journal of Materials Science: Materials in Electronics, 25*(2), 888–896.

Lazenka, V., Lorenz, M., Modarresi, H., et al. (2015). Magnetic spin structure and magnetoelectric coupling in BiFeO$_3$-BaTiO$_3$ multilayer. *Applied Physics Letters, 106*(8), 082904.

Lopez-Varo, P., Bertoluzzi, L., Bisquert, J., et al. (2016). Physical aspects of ferroelectric semiconductors for photovoltaic solar energy conversion. *Physics Reports, 653*, 1–40.

Matsukura, F., Tokura, Y., & Ohno, H. (2015). Control of magnetism by electric fields. *Nature Nanotechnology, 10*(3), 209–220.

Mocherla, P. S., Karthik, C., Ubic, R. N. A. R., Ramachandra Rao, M. S., & Sudakar, C. (2013). Tunable bandgap in BiFeO$_3$ nanoparticles: The role of microstrain and oxygen defects. *Applied Physics Letters, 103*(2), 022910.

Mohanty, B., Bhattacharjee, S., Nayak, N. C., Parida, R. K., & Parida, B. N. (2021). Dielectric, magnetic and optical study of La-doped BFO-BST ceramic for multifunctional applications. *Materials Science in Semiconductor Processing, 128*, 105720.

Moharana, S., Mishra, M. K., Chopkar, M., & Mahaling, R. N. (2016). Enhanced dielectric properties of surface hydroxylated bismuth ferrite—Poly (vinylidene fluoride-co-hexafluoropropylene) composites for energy storage devices. *Journal of Science: Advanced Materials and Devices, 1*(4), 461–467.

Muneeswaran, M., Jegatheesan, P., & Giridharan, N. V. (2013). Synthesis of nanosized BiFeO$_3$ powders by co-precipitation method. *Journal of Experimental Nanoscience, 8*(3), 341–346.

Najm, A. A. A., Baqiah, H., Shaari, A. H., Kechik, M. M. A., Kien, C. S., Zahari, R. M., & Li, Q. (2021). Investigation of structural, dielectric, impedance and magnetic properties of BiFe1-xInxO3 (0.0≤ x≤ 0.6) ceramics. *Results in Physics, 28*, 104550.

Nechache, R., Harnagea, C., Li, S., Cardenas, L., Huang, W., Chakrabartty, J., & Rosei, F. (2015). Bandgap tuning of multiferroic oxide solar cells. *Nature Photonics, 9*(1), 61–67.

Paillard, C., Bai, X., Infante, I. C., et al. (2016). Photovoltaics with ferroelectrics: Current status and beyond. *Advanced Materials*, *28*(26), 5153–5168.

Qi, J., Zhang, Y., Wang, Y., et al. (2017). Effect of Cr doping on the phase structure, surface appearance and magnetic property of BiFeO₃ thin films prepared via sol—gel technology. *Journal of Materials Science: Materials in Electronics*, *28*(23), 17490–17498.

Rahman, M. U., Qazi, U. Y., Hussain, T., Nadeem, N., Zahid, M., Bhatti, H. N., & Shahid, I. (2021). Solar driven photocatalytic degradation potential of novel graphitic carbon nitride based nano zero-valent iron doped bismuth ferrite ternary composite. *Optical Materials*, *120*, 111408.

Ramesh, R., & Spaldin, N. A. (2007). Multiferroics: Progress and prospects in thin films. *Nature Materials*, *6*(1), 21–29.

Remya, K. P., Prabhu, D., Joseyphus, R. J., Bose, A. C., Viswanathan, C., & Ponpandian, N. (2020). Tailoring the morphology and size of perovskite BiFeO₃ nanostructures for enhanced magnetic and electrical properties. *Materials & Design*, *192*, 108694.

Ren, Y., Nan, F., You, L., et al. (2017). Enhanced photoelectrochemical performance in reduced graphene oxide/BiFeO₃ heterostructures. *Small*, *13*(16), 1603457.

Sahni, M., Mukhopadhyay, S., Mehra, R. M., et al. (2021). Effect of Yb/Co co-dopants on surface chemical bonding states of BiFeO₃ nanoparticles with promising photocatalytic performance in dye degradation. *Journal of Physics and Chemistry of Solids*, *152*, 109926.

Sakar, M., & Balakumar, S. (2017). A mechanistic view into the morphology-reconstruction mediated facile synthesis of bismuth ferrite (BiFeO₃) hierarchical nanostructures. *Nano-Structures & Nano-Objects*, *12*, 188–193.

Shankar, S., Kumar, M., Ghosh, A. K., & Thakur, O. P. (2014). Conduction mechanism and dielectric properties of BiFeO₃—BaTiO₃ solid solutions. *Journal of Materials Science: Materials in Electronics*, *25*(11), 4896–4901.

Sharma, S., Kumar, M., Siqueiros, J. M., & Herrera, O. R. (2020). Phase evolution, magnetic study and evidence of spin-two phonon coupling in Ca modified Bi0.80La0.20FeO3 ceramics. *Journal of Alloys and Compounds*, *827*, 154223.

Sharma, S., Tomar, M., Kumar, A., Puri, N. K., & Gupta, V. (2015). Enhanced ferroelectric photovoltaic response of BiFeO₃/BaTiO3 multilayered structure. *Journal of Applied Physics*, *118*(7), 074103.

Singh, A., Khan, Z. R., Vilarinho, P. M., Gupta, V., & Katiyar, R. S. (2014). Influence of thickness on optical and structural properties of BiFeO₃ thin films: PLD grown. *Materials Research Bulletin*, *49*, 531–536.

Spaldin, N. A., & Ramesh, R. (2019). Advances in magnetoelectric multiferroics. *Nature Materials*, *18*(3), 203–212.

Tan, L. Z., Zheng, F., Young, S. M., Wang, F., Liu, S., & Rappe, A. M. (2016). Shift current bulk photovoltaic effect in polar materials—Hybrid and oxide perovskites and beyond. *NPJ Computational Materials*, *2*(1), 1–12.

Tokura, Y., Kawasaki, M., & Nagaosa, N. (2017). Emergent functions of quantum materials. *Nature Physics*, *13*(11), 1056–1068.

Tu, C. S., Hung, C. M., Xu, Z. R., et al. (2013). Calcium-doping effects on photovoltaic response and structure in multiferroic BiFeO₃ ceramics. *Journal of Applied Physics*, *114*(12), 124105.

Tu, C. S., Xu, Z. R., Schmidt, V. H., Chan, T. S., Chien, R. R., & Son, H. (2015). A-site strontium doping effects on structure, magnetic, and photovoltaic properties of (Bi1− xSrx) FeO3−δ multiferroic ceramics. *Ceramics International*, *41*(7), 8417–8424.

Wu, J., Fan, Z., Xiao, D., Zhu, J., & Wang, J. (2016). Multiferroic bismuth ferrite-based materials for multifunctional applications: Ceramic bulks, thin films and nanostructures. *Progress in Materials Science*, *84*, 335–402.

Yang, M. M., Kim, D. J., & Alexe, M. (2018). Flexo-photovoltaic effect. *Science*, *360*(6391), 904–907.

Yang, X., Su, X., Shen, M., et al. (2012). Enhancement of photocurrent in ferroelectric films via the incorporation of narrow bandgap nanoparticles. *Advanced Materials*, *24*(9), 1202–1208.

Yang, Z., Rajagopal, A., & Jen, A. K. Y. (2017). Ideal bandgap organic—Inorganic hybrid perovskite solar cells. *Advanced Materials*, *29*(47), 1704418.

Yuan, Y., Xiao, Z., Yang, B., & Huang, J. (2014). Arising applications of ferroelectric materials in photovoltaic devices. *Journal of Materials Chemistry A*, *2*(17), 6027–6041.

Zhang, Q., Xu, F., Xu, M., et al. (2017). Lead-free perovskite ferroelectric thin films with narrow direct band gap suitable for solar cell applications. *Materials Research Bulletin*, *95*, 56–60.

Zhang, Y., Sun, H., Yang, C., Su, H., & Liu, X. (2019). Modulating photovoltaic conversion efficiency of BiFeO$_3$-based ferroelectric films by the introduction of electron transport layers. *ACS Applied Energy Materials*, *2*(8), 5540–5546.

10 Sustainable Green Composites for Packaging Applications

Tanika Gupta, K. Pratibha, Tannu Garg, S. Shankar, S. Gaurav, and Rohit Verma

CONTENTS

10.1 INTRODUCTION

For many decades, the focus of the scientific community has been on the exploration of smart composites that are healthy for our environment. Researchers worldwide have been doing rigorous work to produce sustainable, environment-friendly materials that are bio-degradable, cheap, durable, and non-polluting. An important class of such materials is the green composites that are completely eco-friendly in the sense that these are bio-degradable and are formed from renewable resources and have plant or animal-based origin. This area of research has been receiving a lot of attention all around the world (Moustafa et al., 2019). Their usage ranges from household materials, aerospace, building materials and constructions, automotive industries, to biomedical and packaging applications. The various examples of the natural fibers include coir, banana, cotton, bamboo, pineapple, flax, jute, sisal, bagasse, abaca, kenaf, hemp, etc., which are obtained from fruits, seeds, stem, and leaves of the plants (Figure 10.1).

DOI: 10.1201/9781003272625-10

FIGURE 10.1 Natural fiber plants: (a) cotton; (b) sisal; (c) jute; (d) flax; (e) hemp; (f) bamboo; (g) banana; (h) coir; and (i) sugar cane.

Source: Syduzzaman et al. (2020)

Apart from the classification of such fibers, another very important aspect is the production of green composites through composite manufacturing techniques. Some of the techniques are injection moulding, compression moulding, thermoforming, pultrusion, and many more. The size of the composite, cost, and quality are also studied in detail before investing both time and energy (Russell, 2014). Mechanical properties of the green composites such as tensile strength, length, density, and stiffness are some of the significant factors taken in account while doing an in-depth comparison of natural fibers and synthetic fibers. Such modifications can be made to induce necessary changes so as to improve certain properties in the composite. Such a procedure is duly followed by the material scientists while studying smart materials (Mann et al., 2018).

The scope of this chapter extends to the packaging applications of green composites. For many years, fossils fuels such as petroleum were being used to produce packaging materials such as plastic bags, bottles, containers, etc. Studies over the decade have already established the shortcomings of plastic as being a non-biodegradable material, so it is highly necessary for the replacement of the plastic packaging with green composites (La Mantia & Morreale, 2011). Not only are these green composites eco-friendly, but they also provide better thermal and mechanical properties in food packaging industries. Food packaging materials such as bamboo fibers are a superior alternative for packaging fruits, liquid, and vegetables (Syduzzaman et al., 2020). The green composites hold an important application in automotive industries as these were brought in use by Henry Ford in 1942. Jute was adopted as a material in door panels of cars in 1996 by Mercedes and was known to reduce the cost of maintenance, installation, and production. Sustainability implies environmental accessibility and commercial viability of the green composites such that these are recyclable, renewable, and biodegradable (Reisch et al., 2013).

10.1.1 Criterion for Packaging Applications

Life-Cycle Assessment (LCA) is a crucial tool to identify prospective smart materials which are healthy for the environment. It also segregates materials that can have adverse effects upon us. Nowadays, LCA is getting attention to evaluate the materials before making heavy investments. It is a scheme to check whether the material is worth its expenditure or not (Smith et al., 2019). Packaging means preservation, storage, transportation, and display of food content (Figure 10.2). The pivotal things to keep in mind while making packaging materials are whether or not it can be kept safe from chemical or physical damage, whether it can be safe during transportation, and whether the consumer will invest in such a type of packaging material or not. A few properties related to packaging are as follows.

- Thermal properties that take into account the behaviour of the packaging on temperature changes, whether it would be brittle or strong enough to hold itself together. An important parameter is the melting point, glass transition temperature, and heat deflection temperature. Glass transition temperature of the packaging material must be kept lower than the freezer temperature to prevent cracking.
- Mechanical properties include tensile strength, fiber volume, fiber weight, tensile modulus, flexural strength, flexural modulus, etc. that are essential.

FIGURE 10.2 Illustration of Life-Cycle Assessment (LCA)

Source: Smith et al. (2019)

- Physical properties such as water absorption, density, moisture content, and thickness are important. Maintaining uniformity and thickness is very important while producing packaging materials.
- Impact resistance takes into account the toughness of the packaging material. Various tests are done in order to check the impact resistance i.e., information about how much pressure the material can absorb.
- Recyclability is the most interesting and important characteristic of green composites packaging materials. Recycling the material is a way to reduce consumption of raw material.

10.2   GREEN COMPOSITES

Natural fibers used to make green composites are obtained from various parts of a plant and are used as raw materials in paper-producing industries, and bio-energy industries. Plant-based fibers also known as bast fibers such as flax, hemp, jute, kenaf, and ramie are extracted from stems of the plants. It has been observed that such fibers provide strength to the stem of the plants and can therefore be made into ropes, paper, and yarn (Figure 10.3) (Torres-Giner, 2017). Chemical composition of natural fibers plays a vital role in characterizing the fiber such as its properties, structure, processability, and appearance (Jadhav et al., 2019). Two or more fibers are constituted to make a composite having properties of all the contributing raw materials. Table 10.1 illustrates some of the physical and chemical properties of bast fibers that are crucial while comparing with synthetic fibers.

Following are some shortcomings of natural fibers that are necessary to overcome in order to make efficient composites.

- The fibers cannot withstand very high temperatures of up to 200 °C.
- These fibers are hydrophilic in nature i.e., are capable of absorbing water.
- It is hard to maintain consistency and uniformity of these fibers since these are naturally grown and are influenced by climatic conditions.

FIGURE 10.3   Green composite sheets with low (left) and high (right) natural fiber contents.

Source: Torres-Giner (2017)

TABLE 10.1

Some Important Bast Fibers and Their Physical and Chemical Properties

Composition and properties	Jute	Flax	Hemp
Cellulose content (%)	59–72	65–87	Under 80%
Hemicellulose content (%)	12–15	Nil	Nil
Lignin (%)	11.8	small	2–3
Crystallinity index	50–55	65–70	66.9
Moisture regains (%)	13.8	12	13
Strength (g/tex)		30–60	23–50

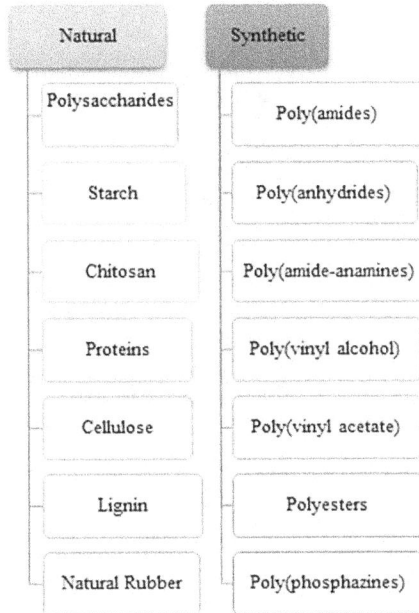

FIGURE 10.4 Classification of biopolymers.

Source: Fabra et al. (2014)

To overcome all these, the natural fibers are often reinforced with plastic. Although these plastics are in low content, such a reinforcement helps to improve the durability of the final composite for practical applications.

The plastics reinforced with natural fibers are thermoplastics such as polypropylene (PP), polyethene, and polyvinyl chloride (PVC) and thermosetting plastics such as epoxy, polyester, and phenolic resin. The reinforced fibers provide benefits like thermal stability, water resistance, chemical resistance, low viscosity, and stress relaxation, economic, storage at room temperature, and reformable. A list of bio-degradable natural and synthetic polymers is shown in the Figure 10.4 (Fabra et al., 2014).

10.2.1 BAMBOO AND ITS APPLICATIONS

Bamboo is the most widely used natural fiber with a percentage usage of about 65%. It is highly abundant in nature so much so that around 30 million tonnes of bamboo fiber is being produced every year (Fazita et al., 2016). India is known to be the single largest producer of bamboo with a sharing of 30% followed by China which is at 14%. Other popular countries producing bamboo include Vietnam, Myanmar, Indonesia, and Ecuador (Figure 10.5).

An attractive attribute associated with bamboo is that it holds excellent tensile strength and is an age-old material known to mankind. Bamboo is comprised of three major chemical compositions contributing to the 95% of bamboo mass. These are cellulose, hemi-cellulose, and lignin (Fazita et al., 2016). Other chemicals in bamboo present in minor amounts are resins, waxes, tannins, and some inorganic salts. The cellulose content in bamboo is around 78.83 percent. These chemicals present in bamboo are known to improve the shelf life of bamboo and protect it from fungus, or mould. Bamboo is a type of grass and can be matured in three to four years. They can grow around 30 inches per day and require low maintenance because they are less susceptible to viruses and fungus infections. Bamboo is an extraordinary material and can be utilized to make utensils such as plates, cups, bottles, etc. (Figure 10.6) (Jabeen et al., 2015).

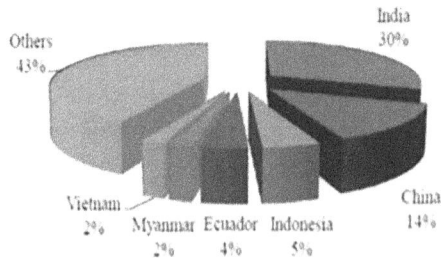

FIGURE 10.5   Countries with largest bamboo resources.

Source: Fazita et al. (2016)

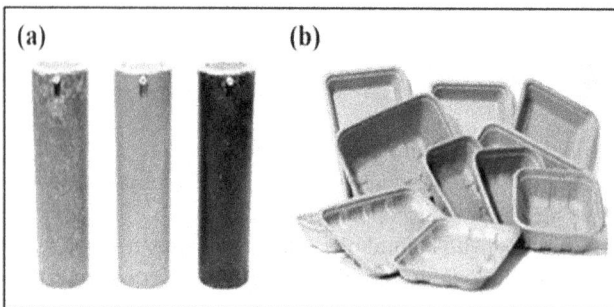

FIGURE 10.6   (a) Cosmetic packaging products made of wood plastic composites and (b) food packaging material made from bamboo fiber.

Source: Jabeen et al. (2015)

The process of making bamboo packaging starts right after its maturity. The matured bamboo is taken for grinding wherein it is made into fine power. This fine powder can be fabricated into different shapes and sizes depending on the manufactures' needs. Bamboo can be disposed after use and is known to reduce carbon footprints by releasing 35% oxygen in the environment (Hai et al., 2020).

Boxes made from bamboo to keep jewellery are both a smart and attractive choice and hold a potential capital since private firms endorsing eco-friendly materials buy these products on a mass scale thereby inducing a greener option in front of the consumers. Apart from these, bamboo fibers are also emerging as they hold peculiarity of anti-bacterial property.

10.2.2 Lignocellulosic Materials

Lignocellulosic materials have the composition as 44% cellulose, 30% hemicellulose, and 26% lignin (Moustafa et al., 2019). The by-products of coffee grounds are known to be lignocellulosic materials. This has come as an attraction to many scientists worldwide as a bio-fuel, or bio-energy materials. Coffee grounds can be realized into green polymers at a pocket friendly cost. Cellulose nanocrystals (CNC) or cellulose nanofibers are lignocellulosic nanofibers that have given commendable results in green food packaging and energy storage applications. These are bio-degradable nanofibers and have more advantage towards inorganic nanofibers in the sense of their low energy consumption, low density, and high mechanical properties (Moustafa et al., 2019). ZnO and CNC films are used to make antimicrobial food packaging containers because of their enhanced shelf-life time (Ghaderi et al., 2014). Coffee ground composites can be used as adsorbents to reduce contamination from waste water (Songtipya et al., 2019).

Cellulose is an organic polymer and is one of the most vital raw materials with a biomass of approximately 10^{12} million tonnes of annual production (Moustafa et al., 2019). Cellulose is found in wood, water plants, agricultural residue, grasses, and algae. Cellulosic units are linked together by glycosidic linkages (Abdul Khalil et al., 2012). Since wood is a primary source of cellulose, cellulose fibers in wood are extracted using mechanical or chemical techniques. Mechanical procedures of acquiring cellulose from wood is an energy consuming technique. In mechanical pulping, wood pulp is attained by rotating wood logs around a cylindrical chamber made of sandstone. However, the pulp produced by this method is homogenized by using refiners (Abdul Khalil et al., 2012).

10.2.3 Bio-Polymers

Polymers that are extracted from renewable resources are known as bio-polymers (Fabra et al., 2014). Bio-polymers can be classified as natural as well as synthetic. Natural bio-polymers are made of starch, cellulose, agar, and soy protein, while synthetic bio-polymers contain polyvinyl alcohol (PVA), polylactic acid (PLA), polybutylene adipate-co-terephthalate (PBAT) etc. Certain drawbacks associated with such fibers can be overcome by reinforcement techniques such as Montmorillonite (MMT) and layered double hydroxide (LDH) coating technique. These reinforcements can drastically improve certain properties of bio-polymers such as mechanical, thermal, and flame retardant properties. Bio-polymers are 100% bio-degradable and this bio-degradability is achieved using bio-active media such as algae, fungi, or by humus. Enzyme action on bio-degradable polymers helps them to decompose in the environment without inducing any damage to it (Dixit & Yadav, 2020).

10.2.4 OIL PALM MICROFIBER-REINFORCED HAND SHEET-MOULDED THERMOPLASTIC

Oil palm empty fruit bunch (OPEFB) is a natural fiber most commonly used in Malaysia. Malaysia is an important producer of palm oil with a production amount of around 18 million tonnes every year (Hermawan et al., 2019). The people of Malaysia are heavily reliant of palm oil cultivation industries. OPEFB is a natural fiber very commonly used as a fertilizer or a biofuel (Figure 10.7). The introduction of OPEFB with a plastic helps in reducing environmental waste and can prove to be a beneficial microfiber. It is a tedious work to synchronize the properties of a natural fiber such as OPEFB with a thermoplastic since the natural fibers can absorb water resulting in decrease in some of the important mechanical properties. Studies suggest that on reinforcement of OPEFB pulp as a filler with polyester matrix, the resulting microfiber shows better performance as packaging materials such as improved tensile strength, elongation at break, flexural strength, etc. (Hermawan et al., 2019).

10.2.5 SOY PROTEIN ISOLATE (SPI)

The proteins and their types are known to be used in manufacturing green composites due to the fact that polysaccharides provide a barrier against O_2 and CO_2 which is an essential requirement of food packaging products (Koshy et al., 2015). There can be a diverse range of protein-based green composites created and used as packaging materials. Soy protein green composites are an interesting area of research for scientists due to its abundant functional properties and sustainability in nature. Other materials

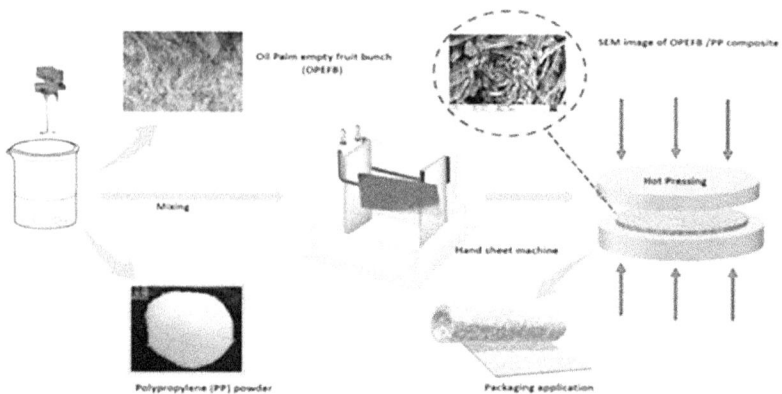

FIGURE 10.7 Life-cycle production of OPEFB into packaging applications.

Source: Hermawan et al. (2019)

where soy protein is used are hydrogels, adhesives, emulsifiers, plastics, etc. Soy protein used in making packaging materials holds great mechanical properties. It can be extracted from soybean seeds and meets requisite food packing standards. To further enhance the properties of SPI, it is reinforced with cellulose and is a promising contribution in biomedical applications (Jadhav et al., 2019).

10.2.6 NANO-COMPOSITE MATERIALS

Chitosan is a type of sugar extracted from the exoskeleton of crab, shrimp, and lobster. It is used as a coating for edible food packaging due to active amino and hydroxyl functional groups. The functional properties of chitosan can be improved by blending the mix with other polymers such as guar gum (Ramadan et al., 2020). Studies have been made to improvise by adding an extract of conocarpus erectus and date seed extract in order to improve anti-microbial properties. The preparation of date seed extract starts by drying seeds in a microwave oven and then grinding them in a coffee machine.

Chitosan slurry is prepared in a conical flask by adding 4% sodium hypophosphate with 5% citric acid and 2% chitosan. Chitosan and date seed extract is being used to treat cotton fiber and realize it into a functional packaging material. This cotton fiber is reinforced with $AgNO_3$ nanoparticles. Each step involved is carefully executed while making a composite including different functional properties of individual elements being brought together into one single material. The material thus formed is carried out for certain measurements such as scanning electron microscopy or SEM to study the internal structure of the fiber obtained. Agar plate method is a method of checking anti-microbial properties of the fabric samples followed by XRD measurements. These measurement techniques are important to characterize the material before handing it on for industry production. The results obtained are critically analyzed and compared with already present data. Other valuable measurements are water absorbency, tensile strength at elongation break, and air permeability. In conclusion, cotton fiber blended with silver nanoparticles, chitosan, and date seed extract shows commendable results as a potential composite for food packaging applications (Jayaraj et al., 2020). Figure 10.8 depicts natural sources based on green composite materials (Mansor et al., 2020).

10.3 FOOD COATING AND EDIBLE FILMS FOR FOOD PRODUCTS

Edible food coatings are prepared from natural and bio-degradable polymers in order to preserve food for a longer duration of time by increasing its shelf life (Reisch et al., 2013). These food coatings have been in use for many decades to prevent moisture, gas, or lipid migration; maintain appearance; and extend shelf life. Glycerol, polyethylene glycol, and acetylated monoglycerides are plasticizers that are used to modify the functional properties of the food coatings and films (Guazzotti et al., 2014). For example, hydrophilic plasticizers are used to improve water permeability of the film (Fabra et al., 2014). A superior advantage of such film coatings is that these induce antioxidants and anti-microbial agents, or functional ingredients such as vitamins, minerals, or probiotics into the food. This makes food safe to consume while also enhancing nutritional values. Edible films can also be used for incorporating aroma and flavour in the food. Studies have been made to modify edible film coatings with

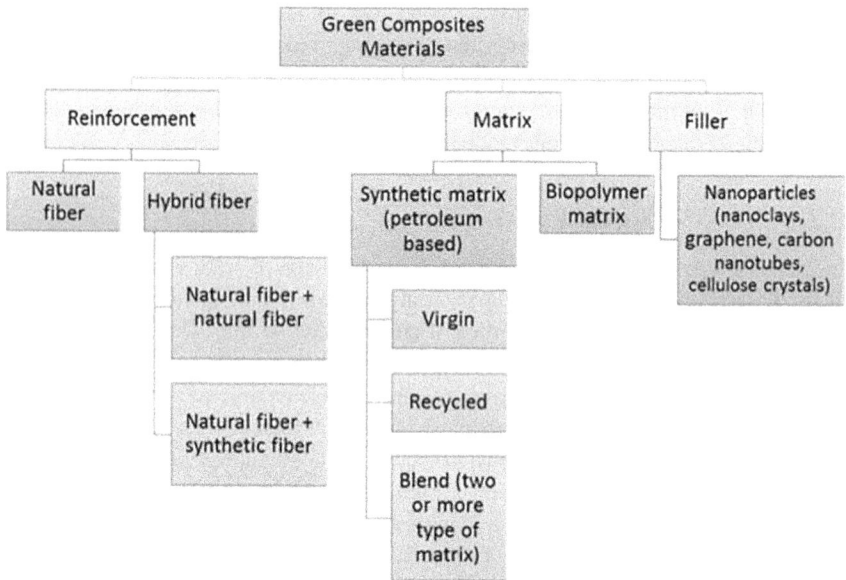

FIGURE 10.8   Natural resources-based green composite materials.

Source: Mansor et al. (2020)

nanoparticles that are healthy for the food as well as for the people who consume it. An ideal film coating has many attributes such as

- No toxicity or allergic substances should be present in them.
- These film coatings should have adhesion with food contents.
- They should be economic, low cost, and easy to manufacture.
- The should be able to enhance sensory attributes of the food contents.
- They should be semipermeable so as to maintain internal equilibrium including aerobic and anaerobic respiration.
- They should be stable both structurally and chemically so that there is no damage during transportation of food contents.
- The film coatings must prevent loss of aroma, flavours, nutritional value, and organoleptic characteristics as well.

If most or all of the above attributes are met, then the edible film coating can be considered as successful for practical use. The food product is either dipped into the edible food coating or it gets sprayed on. However, it is necessary to optimize the concentration of such edible food coatings and note that how much of them should be used on the food contents so as to maintain balance. Coatings provide a protective layer on the food particles and thus maintain structural integrity (Figure 10.9). Coatings made up of a single polymer are prone to crack and are brittle. Any crack in the food coating can cause food damage. To improve this, plasticizer-polymer ratio is calculated to find out the functional properties of the coating material. To prevent harm to perishable foods,

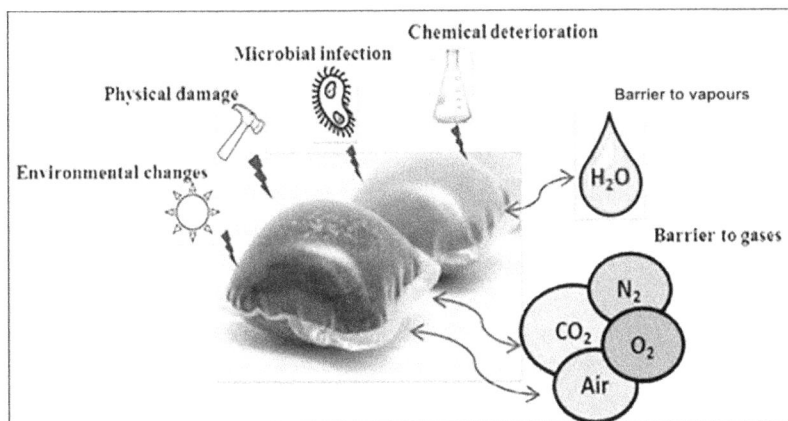

FIGURE 10.9 Functions and usefulness of edible food coatings and films in food packaging.

Source: Hasan et al. (2020)

active food coatings are used containing anti-microbial properties to inhibit the growth of microorganisms (Hasan et al., 2020).

Chitosan is known as an anti-microbial agent which when applied on salmon fillet pieces causes anti-microbial action and prevents damage. A frequently used anti-microbial enzyme is known as lysozyme used in packaging materials and it is beneficial against bacteria. Nisin produced by Lactococcus lactis is a polypeptide also used in the preparation of edible food coatings to prevent bacteria. Butylated hydroxytoluene is a synthetic antioxidant used to make food packaging films and coatings. While on the subject of natural substances, alpha-tocopherol, ascorbic acid, and citric acid are also used as natural antioxidants (Sanches-Silva et al., 2014). Alpha-tocopherol when mixed with xanthan gum improved the nutrition quality of baby carrots. A few applications of edible film coatings in processed foods are

- Prevent moisture loss
- Prevent ice formation in frozen food
- Prevent water vapour transfer between different food contents in heterogenous food system
- Prevent diffusion of carbon dioxide

Edible food coatings are also used for meat and seafood and are known to enhance their quality. Fish fillets when coated with edible food coating films can be stored as long as seven months at a temperature of $-18°C$. These coatings can also be applied on vegetables or fresh cut fruits.

10.4 FOOD PACKAGING

Starch, Poly-Lactic acids (PLAs), and Poly-hydroxy alkanoates (PHAs) are some of the widely used thermoplastics that are sustainable for green food packaging materials. Out of these, PLA and starch are biodegradable polymers and are abundantly available for commercial requirements. PLA provides excellent properties such as transparency and water resistance. Shortcomings related to PLA include low thermal properties, low water permeability, brittleness, and barrier to oxygen and water, as compared to other packing polymers such as PETs. Major properties required while dealing with such materials are

- High chemical resistance
- Easy to process, manufacture, and commercialize
- Barrier to gases and moisture
- Perm selectivity
- Easy to recycle
- Biodegrade

Scientists all around the world are sincerely putting in efforts to develop technologies that are both smart and practical while producing sustainable green composites for packaging applications. The various technological advancements include

- To create multi-layered packages including polymeric materials that can be synthesized by lamination, and co-injection
- To prepare aluminium metallized polymeric films and coatings made by vacuum deposition techniques
- Aluminium and silicon oxides coated with polymeric films

Multi-layered systems provide excellent mechanical, thermal, optical, and processing properties along with high barrier to moisture content. 'Oxygen-scavenger' is a technology wherein we enable low level contact of oxygen with the food products by trapping oxygen in between the headspace and outside of the package.

PLA is known to be the most sustainable alternative till date to its counterpart PET. Other useful technologies are nano bio-composites to improvise barrier properties to permanent gases such as O_2 and CO_2. As already discussed earlier, nano-materials being minute in dimensions can enter the food contents and can be harmful if not made food friendly. Electrospinning is a technique in which ultrafine films of size ranging from 50–500 cm^3 provide enhanced stability, lightness, flexibility, and good mechanical properties. It is therefore a satisfactory method to generate ultrathin fibers made from biodegrade materials. A variety of nanofibers are being realized using the electrospinning technique (Ozdemir & Floros, 2004).

10.5 CHALLENGES INVOLVED

Challenges to food packaging need to be addressed and for that, a concept of intelligent food packaging is followed by industries. A green composite packaging material will be considered intelligent only if it provides usefulness both externally as well as internally (Figure 10.10) (Veisi et al., 2012). The information about the food package is given to the customer by barcode labels, bio-sensors, and time temperature indicators to detect and record any internal or external changes in the food contained.

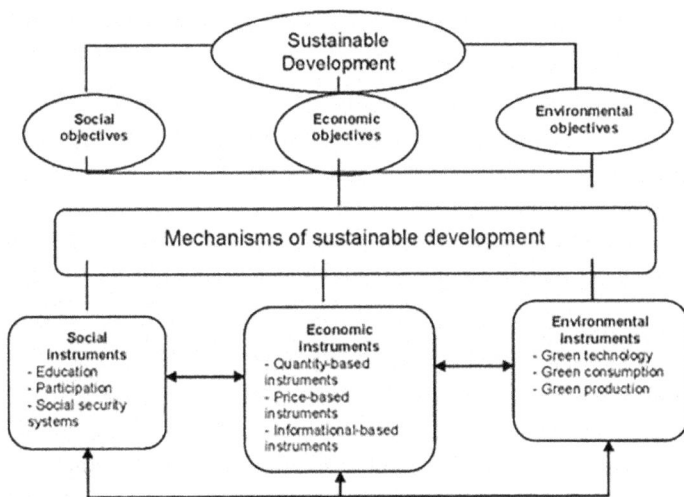

FIGURE 10.10 Concept of mechanism of sustainable development.

Source: Veisi et al. (2012)

A time-temperature indicator is a tool which depicts the time duration for which the food items will remain healthy to eat at a particular temperature.

Despite putting up these intelligent indicators on the food packages, it is still a challenge to increase the shelf life of food items due to microbial action of the food container. Hence, a smart composite that has anti-microbial properties has been in the works for many years now. Metal/metal oxides, organoclay, or some natural materials rosins are used to enhance the anti-microbial properties of the food packaging materials. Nano-zinc and nano-silver nanoparticles have a tremendous potential for inducing anti-microbial properties in food packaging containers and provide protection against fungi and bacteria. It is tricky to manufacture food packages made with nanoparticles because nanoparticles being so minute in size can get into food contents. The food can be subjected to contamination by nanoparticles and it is crucial to make eco-friendly packaging as advised by WHO standards (Grönman et al., 2013).

The raw materials used in preparing green composites are so chosen such that these biodegrade both in fresh as well as marine water, although some studies suggest that biodegradability was better observed in fresh water such as river Nile than in marine water because of the presence of algae in fresh water that could attach polymeric chains and help in decomposition.

10.6 CONCLUSION

This chapter focuses on green composites or the bio-composites that are produced from natural resources that are bio-degradable. Various green fibers such as bamboo fibers, cellulose fibers, soy-protein fibers, lignocellulosic fibers, etc., have been thoroughly discussed along with their potential applications. Edible food coating and packaging films have also been discussed, along with a follow-up scope of the same. There is a need to provide ecological balance by incorporating technologies that are healthy for the environment. However, to enhance durability and life of such materials, one has to keep in check the techniques involved in material synthesis. Certain characteristics like heat capacity, thermal conductivity, and crystallinity of the material provide us insight into the nature of the material, and whether or not to invest in it or not (Mann et al., 2018).

As a follow-up to this area of research, scientists and organizations must ensure extraction of fibers under standardized methods that provide efficiency in the composite matrix and prevent fiber variability; new innovative pathways of composite processing to overcome a few failures such as moisture absorption, humidity, exposure to temperature of UV radiations, and aging due to different external stimuli; and funding in production of green-composites at the school and university level to create awareness of the young budding generation towards the environment's vulnerability towards petroleum-based composites. The residues of fibers must be examined in order to understand their chemical compositions and what steps can be taken to decrease pollution caused by such residues.

REFERENCES

Abdul Khalil, H. P. S., Bhat, A. H., & Ireana Yusra, A. F. (2012). Green composites from sustainable cellulose nanofibrils: A review. *Carbohydrate Polymers*, *87*(2), 963–979. https://doi.org/10.1016/J.CARBPOL.2011.08.078

Dixit, S., & Yadav, V. L. (2020). Biodegradable polymer composite films for green packaging applications. In *Handbook of Nanomaterials and Nanocomposites for Energy and Environmental Applications* (pp. 1–17). https://doi.org/10.1007/978-3-030-11155-7_157-1

Fabra, M. J., López-Rubio, A., & Lagaron, J. M. (2014). Biopolymers for food packaging applications. *Smart Polymers and Their Applications*, 476–509. https://doi.org/10.1533/9780857097026.2.476

Fazita, M. R. N., Jayaraman, K., Bhattacharyya, D., Haafiz, M. K. M., Saurabh, C. K., Hussin, M. H., & Abdul Khalil, H. P. S. (2016). Green composites made of bamboo fabric and poly (Lactic) acid for packaging applications—A review. *Materials*, *9*(6), 435. https://doi.org/10.3390/MA9060435

Ghaderi, M., Mousavi, M., Yousefi, H., & Labbafi, M. (2014). All-cellulose nanocomposite film made from bagasse cellulose nanofibers for food packaging application. *Carbohydrate Polymers*, *104*(1), 59–65. https://doi.org/10.1016/J.CARBPOL.2014.01.013

Grönman, K., Soukka, R., Järvi-Kääriäinen, T., Katajajuuri, J.-M., Kuisma, M., Koivupuro, H.-K., Ollila, M., Pitkänen, M., Miettinen, O., Silvenius, F., Thun, R., Wessman, H., & Linnanen, L. (2013). Framework for sustainable food packaging design. *Packaging Technology and Science*, *26*(4), 187–200. https://doi.org/10.1002/PTS.1971

Guazzotti, V., Marti, A., Piergiovanni, L., & Limbo, S. (2014). Bio-based coatings as potential barriers to chemical contaminants from recycled paper and board for food packaging. *Food Additives and Contaminants—Part A Chemistry, Analysis, Control, Exposure and Risk Assessment*, *31*(3), 402–413. https://doi.org/10.1080/19440049.2013.869360

Hai, L. Van, Choi, E. S., Zhai, L., Panicker, P. S., & Kim, J. (2020). Green nanocomposite made with chitin and bamboo nanofibers and its mechanical, thermal and biodegradable properties for food packaging. *International Journal of Biological Macromolecules*, *144*, 491–499. https://doi.org/10.1016/J.IJBIOMAC.2019.12.124

Hasan, M., Ajesh Kumar V., Maheshwari, C., & Mangraj, S. (2020). Biodegradable and edible film: A counter to plastic pollution. *International Journal of Chemical Studies*, *8*(1), 2242–2245. https://doi.org/10.22271/CHEMI.2020.V8.I1AH.8606

Hermawan, D., Hazwan, C. M., Owolabi, F., Gopakumar, D. A., Hasan, M., Rizal, S., Aprilla, N. S., Mohamed, A., & Khalil, H. A. (2019). Oil palm microfiber-reinforced handsheet-molded thermoplastic green composites for sustainable packaging applications. *Progress in Rubber, Plastics and Recycling Technology*, *35*(4), 173–187. https://doi.org/10.1177/1477760619861984

Jabeen, N., Majid, I., & Nayik, G. A. (2015). Bioplastics and food packaging: A review. *Cogent Food and Agriculture*, *1*(1). https://doi.org/10.1080/23311932.2015.1117749

Jadhav, A. C., Pandit, P., Gayatri, T. N., Chavan, P. P., & Jadhav, N. C. (2019). *Production of Green Composites from Various Sustainable Raw Materials*, 1–24. https://doi.org/10.1007/978-981-13-1969-3_1

Jayaraj, K., Christy, S., & Pius, A. (2020). Green composite film for food packaging applications. *Advances and Applications in Mathematical Sciences*, *20*(2).

Koshy, R. R., Mary, S. K., Thomas, S., & Pothan, L. A. (2015). Environment friendly green composites based on soy protein isolate—A review. *Food Hydrocolloids*, *50*, 174–192. https://doi.org/10.1016/J.FOODHYD.2015.04.023

La Mantia, F. P., & Morreale, M. (2011). Green composites: A brief review. *Composites Part A: Applied Science and Manufacturing*, *42*(6), 579–588. https://doi.org/10.1016/J.COMPOSITESA.2011.01.017

Mann, G. S., Singh, L. P., Kumar, P., & Singh, S. (2018). Green composites: A review of processing technologies and recent applications. *Journal of Thermoplastic Composite Materials*, *33*(8), 1145–1171. https://doi.org/10.1177/0892705718816354

Mansor, M. R., Taufiq, M. J., & Ab Ghani, A. F. (2020). Natural resources based green composite materials. *Composite Materials: Applications in Engineering, Biomedicine and Food Science*, 169–199. https://doi.org/10.1007/978-3-030-45489-0_7

Moustafa, H., Youssef, A. M., Darwish, N. A., & Abou-Kandil, A. I. (2019). Eco-friendly polymer composites for green packaging: Future vision and challenges. *Composites Part B: Engineering*, *172*, 16–25. https://doi.org/10.1016/J.COMPOSITESB.2019.05.048

Ozdemir, M., & Floros, J. D. (2004). Active food packaging technologies. *Critical Reviews in Food Science and Nutrition*, *44*(3), 185–193. https://doi.org/10.1080/10408690490441578

Ramadan, M. A., Sharawy, S., Elbisi, M. K., & Ghosal, K. (2020). Eco-friendly packaging composite fabrics based on in situ synthesized silver nanoparticles (AgNPs) & treatment with Chitosan and/or Date seed extract. *Nano-Structures & Nano-Objects*, *22*, 100425. https://doi.org/10.1016/J.NANOSO.2020.100425

Reisch, L., Eberle, U., & Lorek, S. (2013). Sustainable food consumption: An overview of contemporary issues and policies. *Sustainability: Science, Practice, and Policy*, *9*(2), 7–25. https://doi.org/10.1080/15487733.2013.11908111

Russell, D. A. M. (2014). Sustainable (food) packaging—An overview. *Food Additives & Contaminants: Part A*, *31*(3), 396–401. https://doi.org/10.1080/19440049.2013.856521

Sanches-Silva, A., Costa, D., Albuquerque, T. G., Buonocore, G. G., Ramos, F., Castilho, M. C., Machado, A. V., & Costa, H. S. (2014). Trends in the use of natural antioxidants in active food packaging: A review. *Food Additives and Contaminants—Part A Chemistry, Analysis, Control, Exposure and Risk Assessment*, *31*(3), 374–395. https://doi.org/10.1080/19440049.2013.879215

Smith, L., Ibn-Mohammed, T., Koh, L., & Reaney, I. M. (2019). Life cycle assessment of functional materials and devices: Opportunities, challenges, and current and future trends. *Journal of the American Ceramic Society*, *102*(12), 7037–7064. https://doi.org/10.1111/jace.16712

Songtipya, L., Limchu, T., Phuttharak, S., Songtipya, P., & Kalkornsurapranee, E. (2019). Poly(lactic acid)-based Composites incorporated with spent coffee ground and tea leave for food packaging application: A waste to wealth. *IOP Conference Series: Materials Science and Engineering, 553*(1), 012047. https://doi.org/10.1088/1757-899X/553/1/012047

Syduzzaman, M., Al-Faruque, M. A., Bilisik, K., & Naebe, M. (2020). Plant-based natural fibre reinforced composites: A review on fabrication, properties and applications. *Coatings, 10*(10), 1–34. https://doi.org/10.3390/COATINGS10100973

Torres-Giner, S. (2017). *Development of PHA/fiber-based composites with antimicrobial performance for active food packaging applications.* http://www.actinpak.eu/wp-content/uploads/2018/08/FP1405_GP3_STSM_Report-Torres.pdf

Veisi, H., Liaghati, H., Hashmi, F., & Edizadehi, K. (2012). Mécanismes et instruments du développement durable. *Development in Practice, 22*(3), 385–399. https://doi.org/10.1080/09614524.2012.664624

11 Sustainable Green Composites for Structural Applications and Its Characteristics
A Review

Ankit, Moti Lal Rinawa

CONTENTS

11.1 INTRODUCTION

A large amount of hazardous waste and non-biodegradable plastics is produced each year and leads to dual environmental concerns of land depletion and pollution for the twenty-first century and forthcoming generation (Maraveas, 2020). Plastics take up much more space than metals as a result of their significantly lower density and are non-disposable because of the properties; it takes decades to decompose in dumps/garbage, rendering that area unsuitable for any other purpose biodegradable materials. On the other hand, biodegradable materials take a few months to decompose completely in the organic soil (Gómez & Michel, 2013). According to the report by (UNDP, 2021), India produces 15 million tonnes of non-biodegradable and toxic plastic waste annually, but out of this only one-fourth is disposed. In today's global communities, the manufacturing of environmentally friendly material systems is needed of the hour that may be recycled or disposed of without hurting the environment. Industrial material manufacturers have begun to consider health issues such as neurotoxicity, as well as environmental issues such as the greenhouse effect as well as global climate change, rising carbon dioxide emissions, and depletion of resources, etc. (Wright, n.d.). Polymer composites made from petrochemicals now account for a large portion of waste materials disposed of in landfills. Because of their tailored high specific tensile capabilities and ease of fabrication into desired shapes, polymer composites have piqued the interest of engineers and researchers alike (Dicker et al., 2014). Some issues are inextricably linked to the production and application of polymer-based materials. Recycling plastics is one viable solution to this problem. The majority of these items still end up in landfills due to practical issues with collection and processing economics. It takes several centuries for used plastic materials to decompose, not decades. As a result, the area contributed to landfills is unfit for any other use.

Over the last several years, there has been a greater realization of the need for sustainable development, which has led to the usage of natural fibers as a unique substitute for synthetic fiber using bio-composites. In recent years, there has been a lot of study on natural fiber polymeric composites (Ankit et al., 2021; Ankit & Rinawa, 2021). High usage of petroleum and its products leads to depletion of petroleum resources. To overcome this problem we need biodegradable composites or polymers that can assist humanity towards a sustainable and better future (Shanmugam et al., 2021).

As a result of these factors, traditional structural materials like metals are being phased out of a variety of applications in favour of new lightweight composites. The majority of composites are made up of at least two stages. (1) Polymer composites (like matrix/resin) are frequently hydrocarbon-based polymers; (2) reinforcing composites (like graphite, aramid, or glass fibers) are frequently synthetic fibers (Westman et al., 2010).

The development in the last two to three decades of safer green polymer production processes is also of major interest. Many resins, including soy-protein resin, have been the subject of a substantial investigation to date (Kim & Netravali, 2010a). Some of these natural fiber reinforced green composites (NFRGC) may be recyclable, as they are made using biodegradable resin and fibers (Lodha & Netravali, 2005b; Qiu & Netravali, 2013). In the last decade, the NFRGCs have been perceived to have undergone significant developments. Because of substantial study into the subject of NFRGCs, these materials are presently being explored for numerous engineering and material science applications and need to be updated to comprehend the performance and various qualities related to with the NFRGCs. Overall, the objective of this analysis is to present inspiring ideas and is essential in next-generation natural fiber reinforced composites for structural applications, as well as applications requiring moderately high temperatures.

This work is divided into multiple sections. In Section 11.2 is mechanical properties and applications of NFRGCs. Section 11.3 discusses the structural applications of green composites. Finally, Section 11.4 discusses the foremost concerns and ideas.

11.2 MECHANICAL PROPERTIES AND STRUCTURAL APPLICATIONS OF GREEN COMPOSITES

The mechanical characteristics of the materials are the most important deciding factors in their structural appropriateness. Scientists are actively working to develop green composites which are an economical optimum combination of biodegradable matrix/resins and natural fiber, with properties that nearly resemble those of conventional structural materials. Structural applications of a composite will sustain/absorb collision to minimize the damage. Materials' mechanical qualities are the major deciding factors in their applicability for structural applications. Tensile, flexural, and impact qualities are the most critical. Various automotive firms such as Mercedes-Benz, Toyota, etc., and aerospace firms (such as Blue Origin, SpaceX, and Airbus) have already started the usage of natural fiber composites for manufacturing the vehicle's interior and the aerospace components (Jawaid & Thariq, 2018; Koronis et al., 2013a, 2013b). The ratio of strength and weight of material is very critical while designing the material. Lightweight materials are especially important when it comes to transportation applications like automotive and aerospace. NFRGCs have low density which might aid in enhancing their fuel efficiency (Thoft-Cristensen & Baker, 1982). Similarly, the use of wood plastics in composites makes them perfect for building businesses (Khalid et al., 2021).

In this review, material mechanical characteristics (viz tensile, corrosion resistance, and tolerance to impact) are classified into the following categories: (1) Long-term properties are commonly used in residential constructions and aeroplane interiors. Tensile and flexural characteristics are two examples of long-term material qualities. Although most natural fiber reinforced green composites degrade over time when exposed to hostile environments, even though they are not significant load-bearing buildings, they can function as effective supplementary supports. (2) Short-term structural qualities, on the other hand, account for a material's capacity to tolerate rapid and huge external stresses or shocks' magnitudes. While most buildings/products are intended to tolerate transient physical impacts, some applications are deemed sacrificial or transitory, and as a result, are anticipated or required to outperform in certain circumstances (Mahesh et al., 2021).

In (Yetiş et al., 2020), the effect of NFRGC's high molecular orientation and crystallinity on tensile strength and Young's modulus was studied. In their experiment, they have used Micro Fibrillated Lignocellulose (MFLC) (presented in Lignin) and Polylactic Acid (PLA). Even at low MFLC levels of 0.5 and 1%, the findings showed a considerable rise in PLA tensile characteristics. Results of the experiment showed that tensile strength increased by 37% and Young's modulus increased by 28% than original PLA.

In (Huang et al., 2019), they investigated the effect of Poly Vinyl Alcohol (PVA) as reinforcement on cellulose made from softwood. Initially, softwood pulp was first treated with NaOH and urea before being used to create the RC-SP. Tensile strength rose by 76 MPa and Young's modulus increased by 11.25 GPa. The authors discovered tensile characteristics improved significantly because of interconnection bonding between the hydrogen molecules of the fiber.

In (Komal et al., 2020), researchers stressed the significance of natural fiber reinforced green composites' short-term structural qualities. They used three different techniques for the processing: (1) Direct Injection Moulding; (2) Extrusion Compression Moulding; (3) Extrusion Injection Moulding. They used 4 to 5mm short banana fibers then they injected with PLA moulding.

In (Bayart, 2019), researchers investigated that an increase in quantity of fiber in PLA leads to an increase in tensile strength. Pre-treatment of fibers has been shown in studies to improve their characteristics and/or compatibility with the resin, both of which can improve the natural fiber reinforced green composites qualities.

In (Insung et al., 2019), the authors recommended the usage of trans crystallization which leads to an increase in Young's modulus by 68%. In their experiment, they used banana fiber and chopped/cut them vertically in small 6mm fibers. Later these fibers fused with NaOH + BFSi solution then with PLA resin.

11.3　STRUCTURAL APPLICATIONS OF HYBRID GREEN COMPOSITES

Numerous studies have been carried out by researchers on the development of hybrid/ advanced green composites having high tensile strength and being fully biodegradable with associated fibers. Some of these have the tensile strength of most of the NFRGCs. These composites are widely used in e-glass and as well commercial composites materials. These commercial composites have wide applications in the construction and vehicle sectors and might help promote sustainability.

In some independent studies, flax yarns and resin generated from modified soy protein concentrate were utilized to create advanced green composites. In (Chabba & Netravali, 2005), they developed flax yarns that fused/heated with Soy Protein Composites (SPC), Glutaraldehyde (GA), and PVA. These composites' flexural and tensile properties were then determined. The experiment showed that the longitudinal strength of the composite increased by five times that of PVA + GA + SPC resins. Flax yarns were mixed with an SPC/PVA/GA resin combination that had been procured. To make green composites reinforced with unidirectional flax yarn, a hand lay-up process is used with heat compression. These composites' flexural property and tensile strength were then determined. Flax yarn reinforced composites had a longitudinal tensile strength of 126 MPa, whereas PVA/GA modified SPC resin had a tensile strength that is 5 times higher, demonstrating superior yarn resin bonding and flax yarn tensile strength. Similarly, these composites' Young's modulus was calculated to be 2.24 GPa, which is more than 100% higher than the 1.1 GPa Young's modulus of PVA/ GA modified SPC resin, illustrating the influence of flax yarn stiffness (8.5 GPa). The composites exhibit a 1.18 GPa longitudinal flexural modulus and an 86 MPa flexural strength. In the transverse direction, the composites had a low flexural modulus of 0.06 GPa and flexural strength of 3.6 MPa. The findings demonstrated that flax yarns had a significant influence on flexural properties in the longitudinal direction, but resin had a huge impact in the transverse direction. In the second investigation, flax threads were utilized to reinforce Phytagel J modified SPC, resulting in green composites that are unidirectional. Phytagel J is a polycarboxylic acid-based reactive modifier that forms strong gels via hydrogen bonding. It is produced by bacterial fermentation. Hand lay-up and hot compression were used to test the flexural and tensile parameters of two composite formulations containing 40% and 20% Phytagel, respectively. In composites containing 40% Phytagel J, poor fiber-resin adhesion resulted in increased fiber pull-outs, resulting in worse tensile properties. The authors highlighted high twists of flax fibers, as well as high viscosity and increased shrinkage of the resin, as probable causes of failure (Lodha & Netravali, 2005a).

High strength fibers, exceptional resin characteristics, and strong fiber-resin interfacial adhesion, according to the authors, are the most important aspects in achieving high tensile qualities in advanced green composites. In the research, liquid crystalline cellulose (LCC) fiber/soy protein isolate (SPI) composites were generated after the production of LCC fibers and potassium hydroxide (KOH) solution while under tension was used. According to the authors, raising the LCC fiber content to 65% would result in composites with tensile strength more than 1 GPa and Young's modulus larger than 37 GPa, making them genuinely "advanced" green composites appropriate for major structural or load-bearing applications (Patil & Netravali, 2016).

In (Kim & Netravali, 2010b), they developed a modified sisal fiber with improved/enhanced tensile strength and crystallinity when sisal fibers fused with NaOH were put under strain and then later dried. Their experiment showed that tensile strength increased to 2.8GPa and Young's modulus increased to 3.8GPa.

In (Le Duigou et al., 2019), researchers developed automated fiber placement (AFP) composites using a deposition approach called vacuum-assisted resin transfer moulding (VARTM) and then these continuous glass/polyamide composites were modelled on a commercial 3D printing system. The tensile characteristics achieved were equivalent to that of thermo-compressed composites made of long flax fibers. While additional advances are possible and will be achieved in the future, achievements like these can pave the way for the use of 3D printed natural fiber reinforced green composites for structural applications.

11.4 MAJOR CONCERNS AND IDEAS FOR THE DEVELOPMENT OF GREEN COMPOSITES

In the automotive, civil sector, and aerospace applications, NFRGCs might be used for secondary or even major structural parts. Mechanical properties of NFRGC are impacted when moisture is present on the surface/interior of fiber. However, it is biodegradable and susceptible to biodegradation, especially when exposed to water (Balakrishnan et al., 2016).

The usage of composites in automobiles has been restricted to non-load-bearing side and front panels, but not too crucial structural components. When used to their full potential, NFRGC was found to be an economical and lightweight solution for building components of the vehicle rather than aluminium and steel composites. In the last few decades, construction companies have mostly been dependent on cement/concrete materials due to their longevity and extraordinarily high compressive strength. Cement-based materials, on the other hand, are prone to microcracks at the cement-aggregate contact or on the surface after hydration (Osuská et al., 2019).

In addition, to avoid lightning damage, aeroplane fuselages must be electrically conductive. Although no substantial research has been conducted to date that proposes natural fiber reinforced green composites as conductive to electricity, the potential to build one never declines. In (Mattana et al., 2011), they suggested the non-biodegradable Polyaniline as a filler agent in epoxidized linseed oil. (Silva et al., 2012) suggested the usage of conductive elements like Cu and Ag for coating the natural fibers.

Bamboo fiber compared to other natural fibers has a low density, low cost, high mechanical strength, and a rapid development rate. However, some of its drawbacks include its porous nature, high moisture content, issues with fine and continuous fiber extraction, as well as heat deterioration throughout the manufacturing process (Zakikhani et al., 2014).

From the preceding discussion, controlling and enhancing the mechanical characteristics of green composites is a difficult task. Some difficulties are preventing the widespread applications of natural fiber reinforced green composites for structural purposes (Lau et al., 2018) such as (1) poor durability (Dicker et al., 2014); (2) insufficient thermal stability (Elsabbagh et al., 2017); (3) high moisture absorption (Dicker et al., 2014); (4) inconsistent fiber characteristics as a result of the source (Abdul Khalil et al., 2015); and (5) low fiber or interfacial resin bonding (Luo & Netravali, 1999). It is clear that the variability in the eventual qualities of NFRGCs makes their total performance relatively unpredictable. Extending this research to natural fiber reinforced green composites could improve their applicability as structural materials with high conductivity. These characteristics serve as a benchmark for users to assess their usefulness and manufacturability.

11.5　CONCLUSIONS AND FUTURE SCOPE

The review presented can lead to two major conclusions: (1) NRFGCs are possibly created having a sufficiently enough mechanical strength. They can be substituted with various conventional materials and resins; (2) Properties of green composites like tensile strength and Young's modulus can be improved by combining different composites or by using hybrid NRFGCs. As a result, hybrid NRFGCs might be a great replacement for many presently utilized reasonably high-strength materials. The popularity of NRFGCs opened up many innovative materials design techniques/technologies such as 3D printing of NRFGCs and the use of commercially available green resins, providing a broader perspective to application-based research. These composites offer enticing mechanical properties that can be tweaked. As a consequence, hybrid NFRGCs would be a suitable replacement for several widely utilized reasonably high-strength composites. While there has been considerable research in the creation of advanced green composites over the last few decades, there is little research on short-term structural properties. This work can be extended/improved by performing large-scale studies on structural application and hybrid NRFGCs.

To overcome barriers, further study is needed such as moisture absorption in outdoor applications for long-term stability. Temperature, humidity, and ultraviolet (UV) radiation, in particular, all have an impact on the service.

REFERENCES

Abdul Khalil, H. P. S., Hossain, Md. S., Rosamah, E., Azli, N. A., Saddon, N., Davoudpoura, Y., Islam, Md. N., & Dungani, R. (2015). The role of soil properties and it's interaction towards quality plant fiber: A review. *Renewable and Sustainable Energy Reviews*, *43*, 1006–1015. https://doi.org/10.1016/j.rser.2014.11.099

Ankit, R. M. L. (2021). Sustainable natural bio-composites and its applications. In Agrawal, R., Jain, J. K., Yadav, V. S., Manupati, V. K., & Varela, L. (Eds.), *Recent Advances in Smart Manufacturing and Materials* (pp. 433–439). New York: Springer. https://doi.org/10.1007/978-981-16-3033-0_41

Ankit, R. M. L., Chauhan, P., Suresh, D., Kumar, S., & Santhosh Kumar, R. (2021). A review on mechanical properties of natural fiber reinforced polymer (NFRP) composites. *Materials Today: Proceedings*. https://doi.org/10.1016/J.MATPR.2021.07.275

Balakrishnan, P., John, M. J., Pothen, L., Sreekala, M. S., & Thomas, S. (2016). 12—Natural fibre and polymer matrix composites and their applications in aerospace engineering. In Rana, S., & Fangueiro, R. (Eds.), *Advanced Composite Materials for Aerospace Engineering* (pp. 365–383). London: Woodhead Publishing. https://doi.org/10.1016/B978-0-08-100037-3.00012-2

Bayart, M. (2019). *Élaboration et caractérisation de biocomposites à base d'acide polylactique et de fibres de lin: Compatibilisation interfaciale par dépôt de revêtements à base d'époxy, de dioxyde de titane, de lignine ou de tanin.* https://savoirs.usherbrooke.ca/handle/11143/15772

Chabba, S., & Netravali, A. N. (2005). 'Green' composites Part 2: Characterization of flax yarn and glutaraldehyde/poly(vinyl alcohol) modified soy protein concentrate composites. *Journal of Materials Science, 40*(23), 6275–6282. https://doi.org/10.1007/s10853-005-3143-9

Dicker, M. P. M., Duckworth, P. F., Baker, A. B., Francois, G., Hazzard, M. K., & Weaver, P. M. (2014). Green composites: A review of material attributes and complementary applications. *Composites Part A: Applied Science and Manufacturing, 56,* 280–289. https://doi.org/10.1016/j.compositesa.2013.10.014

Elsabbagh, A., Steuernagel, L., & Ring, J. (2017). Natural Fibre/PA6 composites with flame retardance properties: Extrusion and characterisation. *Composites Part B: Engineering, 108,* 325–333. https://doi.org/10.1016/j.compositesb.2016.10.012

Gómez, E. F., & Michel, F. C. (2013). Biodegradability of conventional and bio-based plastics and natural fiber composites during composting, anaerobic digestion and long-term soil incubation. *Polymer Degradation and Stability, 98*(12), 2583–2591. https://doi.org/10.1016/j.polymdegradstab.2013.09.018

Huang, B., He, H., Liu, H., Wu, W., Ma, Y., & Zhao, Z. (2019). Mechanically strong, heat-resistant, water-induced shape memory poly(vinyl alcohol)/regenerated cellulose biocomposites via a facile co-precipitation method. *Biomacromolecules, 20*(10), 3969–3979. https://doi.org/10.1021/acs.biomac.9b01021

Insung, T., Chaiyut, N., Ksapabutr, B., & Panapoy, M. (2019). Preparation and properties of silane-treated banana fiber/poly(lactic acid) biocomposites. *IOP Conference Series: Materials Science and Engineering, 544,* 012011. https://doi.org/10.1088/1757-899X/544/1/012011

Jawaid, M., & Thariq, M. (2018). *Sustainable Composites for Aerospace Applications.* London: Woodhead Publishing.

Khalid, M. Y., Al Rashid, A., Arif, Z. U., Ahmed, W., Arshad, H., & Zaidi, A. A. (2021). Natural fiber reinforced composites: Sustainable materials for emerging applications. *Results in Engineering, 11,* 100263. https://doi.org/10.1016/j.rineng.2021.100263

Kim, J. T., & Netravali, A. N. (2010a). Effect of protein content in soy protein resins on their interfacial shear strength with ramie fibers. *Journal of Adhesion Science and Technology, 24*(1), 203–215. https://doi.org/10.1163/016942409X12538812532159

Kim, J. T., & Netravali, A. N. (2010b). Mercerization of sisal fibers: Effect of tension on mechanical properties of sisal fiber and fiber-reinforced composites. *Composites Part A: Applied Science and Manufacturing, 41*(9), 1245–1252. https://doi.org/10.1016/j.compositesa.2010.05.007

Komal, U. K., Lila, M. K., & Singh, I. (2020). PLA/banana fiber based sustainable biocomposites: A manufacturing perspective. *Composites Part B: Engineering, 180,* 107535. https://doi.org/10.1016/j.compositesb.2019.107535

Koronis, G., Silva, A., & Fontul, M. (2013a). Green composites: A review of adequate materials for automotive applications. *Composites Part B: Engineering, 44*(1), 120–127. https://doi.org/10.1016/j.compositesb.2012.07.004

Koronis, G., Silva, A., & Fontul, M. (2013b). Green composites: A review of adequate materials for automotive applications. *Composites Part B: Engineering, 44*(1), 120–127. https://doi.org/10.1016/j.compositesb.2012.07.004

Lau, K., Hung, P., Zhu, M.-H., & Hui, D. (2018). Properties of natural fibre composites for structural engineering applications. *Composites Part B: Engineering, 136,* 222–233. https://doi.org/10.1016/j.compositesb.2017.10.038

Le Duigou, A., Barbé, A., Guillou, E., & Castro, M. (2019). 3D printing of continuous flax fibre reinforced biocomposites for structural applications. *Materials & Design, 180,* 107884. https://doi.org/10.1016/j.matdes.2019.107884

Lodha, P., & Netravali, A. N. (2005a). Characterization of phytagel® modified soy protein isolate resin and unidirectional flax yarn reinforced "green" composites. *Polymer Composites, 26*(5), 647–659. https://doi.org/10.1002/pc.20128

Lodha, P., & Netravali, A. N. (2005b). Effect of soy protein isolate resin modifications on their biodegradation in a compost medium. *Polymer Degradation and Stability, 87*(3), 465–477. https://doi.org/10.1016/j.polymdegradstab.2004.09.011

Luo, S., & Netravali, A. N. (1999). Interfacial and mechanical properties of environment-friendly "green" composites made from pineapple fibers and poly(hydroxybutyrate-co-valerate) resin. *Journal of Materials Science, 34*(15), 3709–3719. https://doi.org/10.1023/A:1004659507231

Mahesh, V., Joladarashi, S., & Kulkarni, S. M. (2021). Damage mechanics and energy absorption capabilities of natural fiber reinforced elastomeric based bio composite for sacrificial structural applications. *Defence Technology, 17*(1), 161–176. https://doi.org/10.1016/j.dt.2020.02.013

Maraveas, C. (2020). Environmental sustainability of plastic in agriculture. *Agriculture, 10*(8), 310. https://doi.org/10.3390/agriculture10080310

Mattana, G., Cosseddu, P., Fraboni, B., Malliaras, G. G., Hinestroza, J. P., & Bonfiglio, A. (2011). Organic electronics on natural cotton fibres. *Organic Electronics, 12*(12), 2033–2039. https://doi.org/10.1016/j.orgel.2011.09.001

Osuská, L., Tažký, M., & Hela, R. (2019). High-performance cement composite for architectural elements with elimination of elements with elimination of micro crack. *International Journal of Engineering and Technology.* https://doi.org/10.7763/ijet.2019.v11.1130

Patil, N. V., & Netravali, A. N. (2016). Microfibrillated cellulose-reinforced nonedible starch-based thermoset biocomposites. *Journal of Applied Polymer Science, 133*(45). https://doi.org/10.1002/app.43803

Qiu, K., & Netravali, A. N. (2013). A composting study of membrane-like polyvinyl alcohol based resins and nanocomposites. *Journal of Polymers and the Environment, 21*(3), 658–674. https://doi.org/10.1007/s10924-013-0584-0

Shanmugam, V., Mensah, R. A., Försth, M., Sas, G., Restás, Á., Addy, C., Xu, Q., Jiang, L., Neisiany, R. E., Singha, S., George, G., JosE E, T., Berto, F., Hedenqvist, M. S., Das, O., & Ramakrishna, S. (2021). Circular economy in biocomposite development: State-of-the-art, challenges and emerging trends. *Composites Part C: Open Access, 5*, 100138. https://doi.org/10.1016/j.jcomc.2021.100138

Silva, M. J., Sanches, A. O., Malmonge, L. F., Medeiros, E. S., Rosa, M. F., McMahan, C. M., & Malmonge, J. A. (2012). Conductive nanocomposites based on cellulose nanofibrils coated with polyaniline-DBSA via in situ polymerization. *Macromolecular Symposia, 319*(1), 196–202. https://doi.org/10.1002/masy.201100156

Thoft-Cristensen, P., & Baker, M. J. (1982). *Structural Reliability Theory and Its Applications.* New York: Springer-Verlag. https://doi.org/10.1007/978-3-642-68697-9

UNDP. (2021, October 22). Plastic waste management in India. *UNDP.* www.in.undp.org/content/india/en/home/projects/plastic-waste-management.html

Westman, M. P., Fifield, L. S., Simmons, K. L., Laddha, S., & Kafentzis, T. A. (2010). *Natural Fiber Composites: A Review* (PNNL-19220). Richland, WA: Pacific Northwest National Lab. (PNNL). https://doi.org/10.2172/989448

Wright, L. (n.d.). Plastic warms the planet twice as much as aviation—Here's how to make it climate-friendly. *The Conversation.* http://theconversation.com/plastic-warms-the-planet-twice-as-much-as-aviation-heres-how-to-make-it-climate-friendly-116376

Yetiş, F., Liu, X., Sampson, W. W., & Gong, R. H. (2020). Acetylation of lignin containing microfibrillated cellulose and its reinforcing effect for polylactic acid. *European Polymer Journal, 134*, 109803. https://doi.org/10.1016/j.eurpolymj.2020.109803

Zakikhani, P., Zahari, R., Sultan, M. T. H., & Majid, D. L. (2014). Extraction and preparation of bamboo fibre-reinforced composites. *Materials & Design, 63*, 820–828. https://doi.org/10.1016/j.matdes.2014.06.058

12 Synthesis of Ionic Polymer Metal Composites for Robotic Application

*Shara Khursheed, K. M. Moeed,
and Mohammad Zain Khan*

CONTENTS

DOI: 10.1201/9781003272625-12

12.1 INTRODUCTION

IPMCs are an attractive research area in the field of biomedical, actuators and in robotic applications (Kaasik et al., 2012). They show large deformation in their shape and size on the application of small voltage. They can show all kind of twisting, rolling, bending and other deformation on the applied voltage (Khursheed et al., 2018). Their unique property of working in dry as well as in wet, humid conditions makes them suitable for robotic applications. They can be used in underwater and firefighting robots also. The large deformation on the application of small voltage gives them unique advantage of lower energy conservation but their high cost due to gold or platinum coatings limits their use as shown in Figure 12.1.

12.1.1 IONIC POLYMER METAL COMPOSITES

Nafion and flemion are two forms of IPMCs. Flemion has large ion exchange properties with high stiffness and more water resistance (Çilingir et al., 2008). Flemion has fast bending towards the anode but there is no mechanism of back relaxation while in Nafion-117. The bending speed is faster, as compared to the flemion with the slow relaxation towards cathode. Due to these properties flemion is used in fuel cells while Nafion is ideal for robotic and biomedical applications (Nemat-Nasser & Zamani, 2003). The presence of hydroflouro double bond carbon atom in Nafion structure provides its hydration and strength (Di Noto et al., 2010).

IPMCs are the electro active polymers (Silverman et al., 2020) which also can be used for artificial muscles due to high fracture toughness (Mirfakhrai et al., 2007). The coating of inert material on EAPs makes them suitable for electrical conductance at small voltages. The comparison of EAP with other materials is shown below in Table 12.1. The EAPs are further classified in subgroups as shown in Table 12.2. There are certain merits and demerits of electronic and ionic EAPs and they are listed in a table (Carpi et al., 2011) as shown in Table 12.3.

The ionic EAPs can be further classified as shown in Figure 12.2:

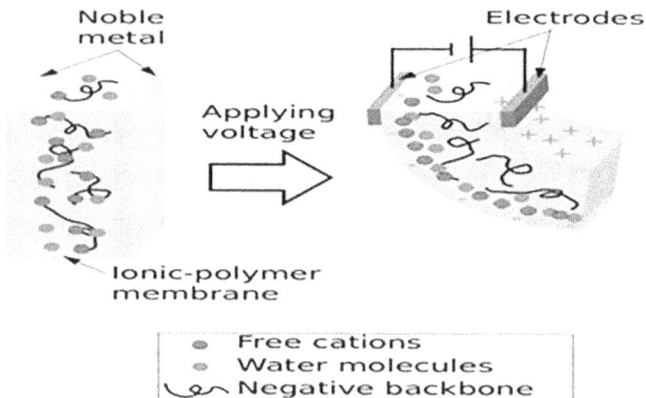

FIGURE 12.1 IPMC working.

TABLE 12.1
Comparison between Smart Materials

Property	Electroactive polymer EAP	Shape memory alloys SMA	Electroactive ceramics EAC
Reaction speed	μ to sec	Sec to min	μ to sec
Actuation Strain	<10%	>8%	1.3%
Force (Mpa)	0.1–0.3	About 700	30–40

Source: L. Chen et al. (2019)

TABLE 12.2
Comparison between EAPs

Electronic EAP	Ionic EAP
Bending due to electric field	Bending due to ion diffusion
Requires high actuation voltage	Low actuation voltage
Only operates in dry conditions	Dry and wet conditions

Source: Breedon & Vloeberghs (2009)

TABLE 12.3
Advantages and Disadvantages of Electronic and Ionic EAPs

EAP	Advantages	Disadvantages
Electronic EAP	Functional at room temperature	Requires high actuation voltage
	Fast response	Actuation direction cannot be changed by voltage polarity.
	Can be activated by DC current	
Ionic EAP	Large bending response on application of small voltage	Cannot produce strain under the DC voltage
	Low actuation voltage	Slow response as compared to electronic EAP
	Bi direction bending	Low electromechanical coupling efficiency

Source: Carpi et al. (2011)

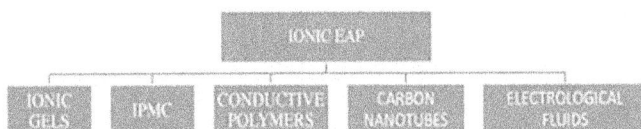

FIGURE 12.2   Classification of IPMC.

12.2   IONIC POLYMER METAL COMPOSITES WORKING PRINCIPLE

IPMCs can be used as actuators or artificial muscles under certain applied voltage (Khursheed et al., 2021). The coated electrodes with gold, platinum or parylene as suggested in this chapter are arranged in a symmetrical or non-symmetrical manner. On the application of voltage if electrodes are arranged in symmetrical manner linear bending is obtained while on arrangement of non-symmetrical electrodes variety of deformation such as bending, twisting and rolling is obtained (Shahinpoor and Kim, 2001). They are electro active polymers which yield output in the term of voltage. They can generate 40 times force of their weight on application of small voltage (Porfiri et al., 2017).

The IPMCs are basically coated with ionomeric membrane, when the voltage is applied across the electrode an electric field induced in the ionomer membrane results in the migration of cations. The migration of cations from the ionomer is responsible for the depletion of electrode and key factor for the actuation of IPMC (Asaka et al., 1995).

12.3   PREPARATION OF IONIC POLYMER METAL COMPOSITES

Earlier the IPMCs were manufactured by electrochemical transduction of polyelectrolyte such as polyvinylchloride (Hamlen et al., 1965). Later on first IPMC was developed in 1965 which has platinum based electrodes and termed as Nafion-115.

IPMCs consist of two parts electrode and a polymer membrane, plated on the electrode by some deposition technique. An IPMC consists of anions, cations and water molecules. Nafion is negatively charged due to the presence of sulfonate group (Bhandari et al., 2012).

12.3.1   Hot Pressing Method

The hot pressing method can be used to produced Nafion based IPMCs or to adhere various films of IPMC together (L. Chen et al., 2019). The IPMC thickness can be easily controlled by this method along with the increase in its stiffness and reproducibility property. The Nafion films are properly dimensioned and heated up to a medium pressure. These films were properly pre-cleaned with the acetone (Marschall et al., 2012). The films are next pressed at 180° Celsius for almost 15–20 mins at atmospheric pressure then a minor pressure of 50MPA for the next 10 mins. These films are next air cooled in hydrogen peroxide solution. The set up was made to measure the tip force of IPMC.

12.3.2   Chemical Deposition

In this the IPMC is prepared by the two steps pre-treatment and impregnation reduction. The powdered form of Nafion is immersed in 160 ml of ammonia solution with 140mg Pd and 120 ml of ammonia for approximately two hours with a slow stirring rate in an ultrasonic environment (30°C–50°C) in the alkaline solution of $NaBH_4$ (2–5%, PH > 13) (Chang & Kim, 2017). This pre-treated membrane is again soaked in ammonia solution for three hours and then electroplating was done on the both side. The electroplated Nafion-117 is immersed in aqueous solution of NaOH (0.1–0.5 mol/L) for a few hours (Zhu et al., 2014).

12.4   IPMC COATING MATERIAL

The coating of IPMC has been done to avoid the loss of water molecules in the humid environment and to provide inert atmosphere to the Nafion-117 electrode. Several materials have been used for the coating of Nafion-117.

12.4.1   PLATINUM

Platinum can be deposited over Nafion-117 by electroless plating. The parameters needed to be controlled are deposition time and deposition rate (Lee et al., 2008) as shown in Table 12.4. The pre-treatment of Nafion is done with deionized water. The interface of platinum Nafion is studied through SEM, TEM, XRD and other techniques.

12.4.2   PALLADIUM

The cost of platinum coating is too high; therefore, an alternative choice for coating that has been developed is palladium. It can be deposited easily by spin coating with the high open circuit voltage (Wagner et al., 2010). This results in better cell or electrode performance at low current density as shown in Table 12.5. The pH values and temperature have to be controlled in order to control the hydrogen deposition.

12.4.3   PARYLENE

This is the polymer of para-xylylene group which exhibit the unique properties that are ideal for coating of smart material. The advantage of parylene coating is that it is

TABLE 12.4
Platinum Coating

Platinum source	Pt conc (g/l)	Main electrolyte (g/l)	Medium	Temp °C
K2Pt(OH)6	10	K_2SO_4	B	70–90
$PtCl_4$	3	Na_2HP_4	B	70–90
H_2PtCl_6	5–25	HCl	A	45–90

Source: Khan et al. (2016)

TABLE 12.5
Palladium Coating

Palladium source	Pt conc (g/l)	Main electrolyte (g/l)	Medium	Temp °C
H_2PdCl_4	05	NaCl	A	25–45
$PdCl_2$	1–40	KH_2PO_4	B	40–70

Source: Wepasnick et al. (2011)

TABLE 12.6
Parylene Types

Property	Parylene C	Parylene D	Parylene N
Melting point °C	420	290	380
Tensile strength (MPa)	45	70	75
Surface resistivity	10^{13}	10^{14}	10^{16}
Water absorption in 24 hrs	0.1	0.1	0.1
Yield strength (MPa)	32	55	60

Source: Lee & Meng (2015)

thin, pin hole free and rust free. The precursor of parylene is paracyclophane and it can be prepared by Hoffmans elimination reaction. The parylene is available in different forms. Parylene N is the basic form of parylene and it can be modified into parylene D or parylene C. The parylene coating can be cheaper as compared to the other material. The SEM, TEM, XRD, CV and ESV test can be done to check the compatibility and strength of parylene coating. In this chapter, parylene coating is suggested for the smart polymer material (Kim & Meng, 2016).

In this chapter, parylene has been selected as coating material to the ionic polymer (IPMC) due to its low cost as compared to other coating materials. The parylene has high wear resistant properties at both elevated and subzero temperature (Ortigoza-Diaz et al., 2018). The properties of parylene types have been compared, and parylene D is selected for coating purpose due to its high stability and corrosion resistant properties shown in Table 12.6.

12.4.4 PLASMA TREATMENT OF PARYLENE

The main problem with the parylene coating is of adhesion with Nafion-117. The parylene D may not show proper adhesion during the elevated temperatures, so in order to overcome this problem the plasma treatment of parylene is done (Schmiedel et al., 2019). The oxygen and sulphur hexafluoride are used for treating parylene. The AFM and X-Ray test were conducted for a check of surface wettability (Banerjee et al., 2011). The results were shown in increased surface roughness, that is increased atomic bonding forces between oxygen, fluoride and the coating surface (Chen et al., 2010).

12.4.5 CHEMICAL VAPOR DEPOSITION

The parylene is coated on Nafion-117 through the chemical vapor deposition process (Wen et al., 2012). In this, the paracyclophane (parylene precursor) is loaded into the sublimation chamber at approximately 120°C, and this parylene is converted into the monomer inside the pyrolysis chamber (650°C). The parylene deposition over the ionic polymer (Nafion-117) at room temperature (25°C) is carried out inside the vaccum. The vaccum was created by the liquid nitrogen or by the cold (Mirfakhrai et al., 2007).

12.5   DETAILED STUDY OF CHEMICAL VAPOR DEPOSITION

12.5.1   SUBLIMATION CHAMBER

Parylene dimer is placed at the end of the sublimation chamber in a tray. The diameter of chamber is approximately 2.0 inches and the material is stainless steel (Gołda et al., 2013). The sublimation chamber should have heating capacity up to 200°C. The sublimation chamber is connected to the pyrolysis chamber directly or through the sublimation tube. The value is generally provided in the tube to regulate the flow of parylene dimer.

12.5.2   PYROLYSIS CHAMBER

The diameter of pyrolysis chamber should be 24–40 inches and its temperature should raise up to 700°C. The pyrolysis has inside linings of quartz material to prevent the corrosion. The thermocouple is also provided to measure the temperature (Senkevich, 2000).

12.5.3   DEPOSITION CHAMBER

The deposition chamber is bell jar shaped in which deposition takes place at room temperature. The vacuum is created inside the deposition chamber. The vacuum is created by the cold trap or the mechanical pump. In Figure 12.3, the pressure measuring unit is attached to the deposition chamber for pressure management (Heyries & Hansen, 2011).

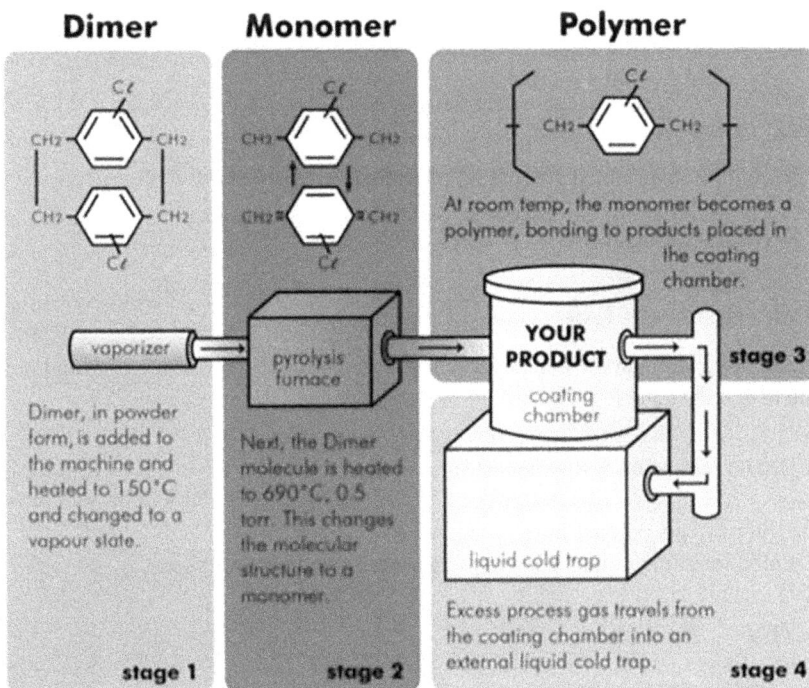

FIGURE 12.3   Chemical Vapor Deposition of Parylene Coating on Nafion-117.

12.6   KINEMATICS OF IONIC POLYMER COATINGS

The kinematics of chemical vapor deposition for ionic polymer composites, the following parameters should be studied.

12.6.1   PUMP SPEED

The measurement of pump speed in the deposition chamber can be done by

1. The base pressure is maintained through the vacuum wall inside the system (Yang et al., 2018). Than the vacuum value is closed, and the pressure rise *dPI I dt* is measured.
2. The base pressure is again maintained by opening the vacuum value and at pressure P_2, gas is admitted till equilibrium.
3. Now, *dP2/dt* is measured in closed valve (Miyakawa et al., 2020).
4. The system volume is estimated or measured.

After these steps have been followed the pumping speed (*S*) can be calculated by using the equation

$$S = V\left\{ \left(\frac{dP2}{dt} - \frac{dP1}{dt} \right) \middle/ (P2 - P1) \right\}$$

Where,
S = pump speed in L/sec.

12.6.2   FLOW RATE

The flow rate is determined at required pressure

$Q = SP$ (Mori et al., 2006).

12.6.3   RESIDENCE TIME

The time of gas flow in the deposition chamber is termed as residence time and it can be calculated through:

$t = L/Vm$ (Arumugam & Reddy, 2018)

where, *t* represents contact time in seconds,
L is tube length in cm, and *Vm* is the gas flow rate in mL/cm²sec

$V_m = D/S$

Where, *D* is total flow rate of gas in ml/sec and *S* is the tube cross sectional area in cm².

12.7   BACK RELAXATION MECHANISM IN IPMC

The IPMC actuation was a result of nonlinear motion of ions. Due to the osmotic pressure there is a rapid bending towards the anode but with principle of Maxwell stress pressure a slight bending also occur towards the cathode (Porfiri et al., 2017).

With the increase in the applied voltage the Maxwell stress also increases and hence results in the increased back relaxation process. To minimize the back relaxation several techniques have been proposed but this phenomenon has been more visible in the case of large bending; however, in small bending it can be controlled to a certain extent (Annabestani et al., 2016). The closed and open loop feedback system has been introduced to minimize this effect. Due to the variable input, the application of closed loop systems is more complex as compared to open loop systems. The patterned electrode model is proposed to eliminate this back relaxation. On the application of voltage nonlinear diffusion of ions towards the anode takes place when this voltage is applied for longer duration then BR effect takes place. it is observed that if flow of water molecules towards the anode is controlled the BR effect will be eliminated (Fleming et al., 2012).

12.8   IONIC POLYMER METAL COMPOSITES AS ARTIFICIAL MUSCLES

The IPMC's unique properties of low actuation voltage and high bending phenomena make it suitable to be used as artificial muscle as shown in Table 12.7.

The use of IPMC is becoming more popular in robots' day by day in replacing the robot motor with the IPMC as artificial muscles which may solve the problematic high operating cost and short life span of robotic motors. The comparison of IPMC with types of robot motors is done next:

12.8.1   IPMC vs Servo Motor

They run on the coding signal given by the shaft and in turn they exhibit linear and rotatory actuation. The lifespan of the motor is only limited to the 2050 hours and after that it's worn out. For higher power output the motor is running at its peak speed, at low speed the power generated by the motor is insufficient.

12.8.2   IPMC vs Stepper Motor

This type of motor cannot run on the fluctuating speed but they have higher lifespans. If it's replaced by the IPMC then they can easily handle the fluctuating load and can provide all types of bending under fluctuating load (Chen et al., 2018).

12.8.3   IPMC vs DC Motor

The DC motor uses a gearbox for the actuation purpose which requires lubrication and other problems. The use of IPMC can cease the limitation associated with the gearbox. The use of gear technology can give the friction problem which is completely eradicated by the use of IPMC.

TABLE 12.7
Comparison between Various Materials as Artificial Muscles

Property	Biological muscles	McKibben muscles	Polyacry lonitrile muscles	Nanotube artificial muscles	IPMC
Strength	They have high strength and elasticity.	They have a modular structure and strain sensors are present.	Strength is of comparison to the biological muscles.	They exhibit very good elastic property.	They are flexible in nature which is clearly indicated by the bending of IPMC when a voltage is applied.
Actuation	The actuation energy is provided by the chemical energy due to the food intake which is then converted to mechanical energy.	The pneumatic stepper motor is used for actuation process. The actuation is obtained by the gas pressure due to the difference between external and internal pressure.	The contraction is by hydrogen ion by the process of electrolysis and generally drives by motor or flywheel.	In this contraction is obtained by the light and in presence of certain chemicals.	They are actuated by the very low voltage that is less than 4 V. The contraction energy is obtained by the flow of anions in the electrolysis.
Weight	The weight of human muscles is 40% of whole body weight.	They are light in weight and easy to manufacture.	They are heavier than the McKibben muscles.	They are less in diameter as compared to the other muscles	They are light in weight with very low density.
Limitation	Human muscles are made of tissues and sometimes due to stress distribution permanent rapture occurs.	Their volume and shape do not change during the contraction, hence their strength is low.	The needs of strong acid and base for their actuation restrict their use.	The dimensions of carbon nanotube muscles of few millimeters make the reinforcement a challenging task.	The only limitation of IPMC is their high coating cost due to the high cost of noble materials.
Conduction	They are a good conductor of heat and electricity.	They are good conductor of electricity.	In order to increase their conductivity platinum coating is used.	They are good conductors of heat and electricity.	In order to increase their conductivity, the various coating materials are used.

Source: Gluschke et al. (2019)

12.9   CONCLUSION AND LIMITATIONS OF IPMC

The use of IPMC is still not very popular in robots due to its high cost and careful handling. During the electrolysis process the ion flow is too much which can weaken the material. In its use as artificial muscle the shape turned on application of pressure is cylindrical which may cause unwanted bending of IPMC.

REFERENCES

Annabestani, M., Maymandi-Nejad, M., & Naghavi, N. (2016). Restraining IPMC back relaxation in large bending displacements: Applying non-feedback local Gaussian disturbance by patterned electrodes. *IEEE Transactions on Electron Devices*, *63*(4), 1689–1695.

Arumugam, J., & Reddy, J. N. (2018). Nonlinear analysis of ionic polymer—Metal composite beams using the von Kármán strains. *International Journal of Non-Linear Mechanics*, *98*, 64–74.

Asaka, K., Oguro, K., Nishimura, Y., Mizuhata, M., & Takenaka, H. (1995). Bending of polyelectrolyte membrane–platinum composites by electric stimuli I. Response characteristics to various waveforms. *Polymer Journal*, *27*(4), 436–440.

Banerjee, I., Pangule, R. C., & Kane, R. S. (2011). Antifouling coatings: recent developments in the design of surfaces that prevent fouling by proteins, bacteria, and marine organisms. *Advanced Materials*, *23*(6), 690–718.

Bhandari, B., Lee, G. Y., & Ahn, S. H. (2012). A review on IPMC material as actuators and sensors: fabrications, characteristics and applications. *International Journal of Precision Engineering and Manufacturing*, *13*(1), 141–163.

Breedon, P., & Vloeberghs, M. (2009). Application of shape memory alloys in facial nerve paralysis. *Australasian Medical Journal*, *1*(11), 129–135.

Carpi, F., Kornbluh, R., Sommer-Larsen, P., & Alici, G. (2011). Electroactive polymer actuators as artificial muscles: Are they ready for bioinspired applications?. *Bioinspiration & Biomimetics*, *6*(4), 045006.

Chang, Y. C., & Kim, W. J. (2017). An electrical model with equivalent elements in a time-variant environment for an ionic-polymer-metal-composite system. *International Journal of Control, Automation and Systems*, *15*(1), 45–53.

Chen, L., Feng, Y., Li, R., Chen, X., & Jiang, H. (2019). Jiles-Atherton based hysteresis identification of shape memory alloy-actuating compliant mechanism via modified particle swarm optimization algorithm. *Complexity*, *2019*, Article ID 7465461. https://doi.org/10.1155/2019/7465461

Chen, S., Li, L., Zhao, C., & Zheng, J. (2010). Surface hydration: Principles and applications toward low-fouling/nonfouling biomaterials. *Polymer*, *51*(23), 5283–5293.

Chen, Z., Hou, P., & Ye, Z. (2018). Modeling of robotic fish propelled by a servo/IPMC hybrid tail. In *2018 IEEE/RSJ International Conference on Intelligent Robots and Systems (IROS)* (pp. 8146–8151). New York: IEEE.

Çilingir, H. D., Menceloglu, Y., & Papila, M. (2008). The effect of IPMC parameters in electromechanical coefficient based on equivalent beam theory. In *Electroactive Polymer Actuators and Devices (EAPAD)* (vol. 6927, p. 69270L). Bellingham, WA: International Society for Optics and Photonics.

Di Noto, V., Negro, E., Sanchez, J. Y., & Iojoiu, C. (2010). Structure-relaxation interplay of a new nanostructured membrane based on tetraethylammonium trifluoromethanesulfonate ionic liquid and neutralized nafion 117 for high-temperature fuel cells. *Journal of the American Chemical Society*, *132*(7), 2183–2195.

Fleming, M. J., Kim, K. J., & Leang, K. K. (2012). Mitigating IPMC back relaxation through feedforward and feedback control of patterned electrodes. *Smart Materials and Structures, 21*(8), 085002.

Gołda, M., Brzychczy-Włoch, M., Faryna, M., Engvall, K., & Kotarba, A. (2013). Oxygen plasma functionalization of parylene C coating for implants surface: Nanotopography and active sites for drug anchoring. *Materials Science and Engineering: C, 33*(7), 4221–4227.

Gluschke, J. G., Richter, F., & Micolich, A. P. (2019). A parylene coating system specifically designed for producing ultra-thin films for nanoscale device applications. *Review of Scientific Instruments, 90*(8), 083901.

Hamlen, R. P., Kent, C. E., & Shafer, S. N. (1965). Electrolytically activated contractile polymer. *Nature, 206*(4989), 1149–1150.

Heyries, K. A., & Hansen, C. L. (2011). Parylene C coating for high-performance replica molding. *Lab on a Chip, 11*(23), 4122–4125.

Kaasik, F., Torop, J., Must, I., Soolo, E., Põldsalu, I., Peikolainen, A. L., & Aabloo, A. (2012, April). Ionic EAP transducers with amorphous nanoporous carbon electrodes. In *Electroactive Polymer Actuators and Devices (EAPAD) 2012* (vol. 8340, p. 83400V). Bellingham, WA: International Society for Optics and Photonics.

Khan, A., & Jain, R. K. (2016). Easy, operable ionic polymer metal composite actuator based on a platinum-coated sulfonated poly (vinyl alcohol)—Polyaniline composite membrane. *Journal of Applied Polymer Science, 133*(33).

Khursheed, S., Chaturvedi, S., & Moeed, K. (2018, December). Comparative study of the use of IPMC as an artificial muscle in robots replacing the motors: A review. *Proceedings of TRIBOINDIA-2018 An International Conference on Tribology.*

Khursheed, S., Khan, M. Z., Moeed, K. M., & Sultana, S. (2021). A review of coating materials for ionic polymer metal compounds for Nafion-117. *Materials Today: Proceedings, 46*(15), 6655–6659.

Kim, B. J., & Meng, E. (2016). Micromachining of parylene C for bioMEMS. *Polymers for Advanced Technologies, 27*(5), 564–576.

Lee, C. D., & Meng, E. (2015). Mechanical properties of thin-film Parylene—Metal—Parylene devices. *Frontiers in Mechanical Engineering, 1*, 10.

Lee, P. C., Han, T. H., Kim, D. O., et al. (2008). In situ formation of platinum nanoparticles in Nafion recast film for catalyst-incorporated ion-exchange membrane in fuel cell applications. *Journal of Membrane Science, 322*(2), 441–445.

Marschall, R., Klaysom, C., Mukherji, A., Wark, M., Lu, G. Q. M., & Wang, L. (2012). Composite proton-conducting polymer membranes for clean hydrogen production with solar light in a simple photo electrochemical compartment cell. *International Journal of Hydrogen Energy, 37*(5), 4012–4017.

Mirfakhrai, T., Madden, J. D., & Baughman, R. H. (2007). Polymer artificial muscles. *Materials Today, 10*(4), 30–38.

Miyakawa, K., Takahama, Y., Kida, K., Sato, K., & Kushida, M. (2020). Fabrication and evaluation of stacked polymer actuator and divided polymer actuator using the electrospinning method. *Japanese Journal of Applied Physics, 59*(SI), SIIF02.

Mori, M., Watanabe, T., Kashima, N., et al. (2006). Development of long YBCO coated conductors by multiple-stage CVD. *Physica C: Superconductivity and its Applications, 445*, 515–520.

Nemat-Nasser, S., & Zamani, S. (2003). Experimental study of Nafion-and Flemion-based ionic polymer metal composites (IPMCs) with ethylene glycol as solvent. In *Smart Structures and Materials 2003: Electroactive Polymer Actuators and Devices (EAPAD)* (vol. 5051, pp. 233–244). Bellingham, WA: International Society for Optics and Photonics.

Ortigoza-Diaz, J., Scholten, K., & Meng, E. (2018). Characterization and modification of adhesion in dry and wet environments in thin-film Parylene systems. *Journal of Microelectromechanical Systems*, *27*(5), 874–885.

Porfiri, M., Leronni, A., & Bardella, L. (2017). An alternative explanation of back-relaxation in ionic polymer metal composites. *Extreme Mechanics Letters*, *13*, 78–83.

Schmiedel, C., Schmiedel, A., & Viöl, W. (2009, July). Combined plasma laser removal of parylene coatings. *ISPC Conference*, 27–31.

Senkevich, J. J. (2000). Thickness effects in ultrathin film chemical vapor deposition polymers. *Journal of Vacuum Science & Technology A: Vacuum, Surfaces, and Films*, *18*(5), 2586–2590.

Shahinpoor, M., & Kim, K. J. (2001). Ionic polymer-metal composites: I. Fundamentals. *Smart materials and structures*, *10*(4), 819.

Silverman, J., Irby, A., & Agerton, T. (2020). Development of the crew dragon ECLSS. *International Conference on Environmental Systems*, 1–11.

Song, J. S., Lee, S., Jung, S. H., Cha, G. C., & Mun, M. S. (2009). Improved biocompatibility of parylene-C films prepared by chemical vapor deposition and the subsequent plasma treatment. *Journal of Applied Polymer Science*, *112*(6), 3677–3685.

Wagner, F. T., Lakshmanan, B., & Mathias, M. F. (2010). Electrochemistry and the future of the automobile. *The Journal of Physical Chemistry Letters*, *1*(14), 2204–2219.

Wen, L., Wouters, K., Ceyssens, F., Witvrouw, A., & Puers, R. (2012). A Parylene temporary packaging technique for MEMS wafer handling. *Sensors and Actuators A: Physical*, *186*, 289–297.

Wepasnick, K. A., Smith, B. A., Schrote, K. E., Wilson, H. K., Diegelmann, S. R., & Fairbrother, D. H. (2011). Surface and structural characterization of multi-walled carbon nanotubes following different oxidative treatments. *Carbon*, *49*(1), 24–36.

Yang, W., Choi, S., Kim, H., Cho, W., & Lee, S. (2018). Theoretical analysis and design for a multilayered ionic polymer metal composite actuator. *Journal of Intelligent Material Systems and Structures, 29*(3), 446–459.

Zhu, Z., Chang, L., Takagi, K., Wang, Y., Chen, H., & Li, D. (2014). Water content criterion for relaxation deformation of Nafion based ionic polymer metal composites doped with alkali cations. *Applied Physics Letters*, *105*(5), 054103.

13 Carbon Footprint Analysis of Green Composites

Tannu Garg, Tanika Gupta, S. Gaurav, S. Shankar, and Rohit Verma

CONTENTS

13.1 INTRODUCTION

As global societies are growing continuously, there has been an increasing emphasis on guaranteeing the sustainability of our materials (Dicker et al., 2014a). Composites are basically a material which has a minimum of two immiscible phases separated by an interface regime, majority of which is called matrix, and fillers are dispersed in it. Composites made up of polymers are widely used in which a polymer with a filler is used to enhance the properties that could be widely used economically. The polymer composites that use polymer and one or more fillers have allowed numerous advantages with mixing the properties of two (Mann et al., 2018). The fillers such as mica, graphite, calcium carbonate, silica and glass fibers, and synthetic fillers such as PET or PVA etc. have provided several advantages. However, the combination of these composites leads to difficulty to reuse or recycle the product. They are either directly disposed into dumps or incinerated, which is very unsatisfactory as well as costly. Also, the biggest challenge is the environmental impact due to these processes (la Mantia & Morreale, 2011).

With the rising concern for environment protection and conservation, there is a need for versatile polymer-based materials which are natural-organic in which the polymer composite fillers are filled with renewable and biodegradable resources. Since the composite made materials are made from renewable resources, they hence are referred to as green-composites (la Mantia & Morreale, 2011). Traditional polymer composites can be substituted with other composite material having lower environmental impact, which can be termed as "green composites" or "eco-composites". The betterment of sustainability in materials requires growth of novel sustainable materials as well as the augmented application of prevailing green materials (Dicker et al., 2014a).

The first attempts to produce green composites was fixated on the synthesis and characterization of polymer composites which is dependent on decomposable polymer which can be filled with natural-organic fillers; these fibers and particles are extracted from plants. There have been many advantages to this, for instance firstly, replacing the traditionally used mineral inorganic fillers with the usage of natural organic fillers has helped in reducing the use of the non-biodegradable polymers and also the use of non-renewable resources. These natural-organic fillers are mainly made from wood flour and fibers (la Mantia & Morreale, 2011). Since the raw materials are plant based (often from waste) and hence are available abundantly with the use of these fillers, they are also very cheap to make with less abrasive than inorganic-mineral counterparts. The green composites tend to involve less machinery, are easy to be incinerated and are also safe for employees helping them in production. All these properties help to produce final composite material which has lower specific weight as compared to their mineral filled counterparts thus having properties like thermal and acoustic insulation (la Mantia & Morreale, 2011).

13.2 GREEN COMPOSITES

Green composites can be described as biopolymers, which are derivates of biopolymers reinforced with natural fibers as shown in Figure 13.1. The green composites are biodegradable (Ead et al., 2021). Due to their renewable nature, these bio-composites have gained a constant increase in applications (Correa et al., 2019). The bio composites have several applications in the automotive industry. These applications are chosen so that the advantages of green composites are maximized for optimum mechanical efficiency (Ead et al., 2021). The main attractive features of these composites are that they are made at a low price, are biodegradable and have high availability, and they also have a capacity to substitute other compounds such as glass or carbon fibers.

The natural fibers that make up the green composite material are obtained from natural resources (Verma et al., 2016). They are limited by low property variation such as fiber shape, length and chemical composition. The crucial parameters that decide the important properties of green fibres are crop variety, fibre location, seed density, soil quality, fertilization, climate and harvest timing. Different attempts and techniques have been implemented to tailor the size, shape and tensile properties through numerical estimation of individual fibre. From studies it has been found that green composites

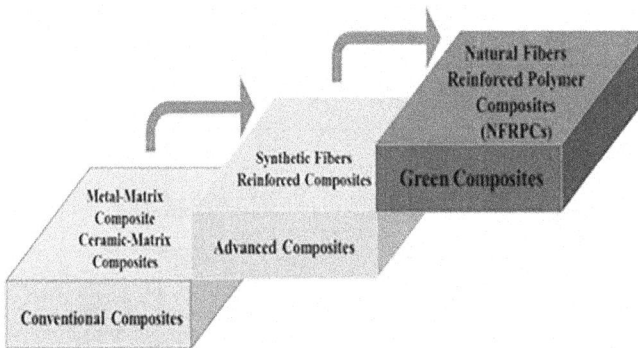

FIGURE 13.1 Different types of composites.

Source: Khalid et al. (2021)

have smaller dimensions of nearly 300 mm in comparison to existing artificial fibers (Koronis et al., 2013). For exploiting the fiber composite properties, maximization of fiber length as well as alignment of fiber along the applied load is necessary. Consequently, improved aligning of fiber permits attainment of large fiber volume fractions. As natural fibers are inexpensive, their manufacturing is dominated by the following parameters (Zini & Scandola, 2011):

1) The treatment and processing methods of green composites for obtaining optimum efficiency rapidly elevates the production cost.
2) The requirement is decided by the functioning of the fiber industry.

13.2.1 NATURAL FIBERS

The natural fibers make green composite materials which are either made from vegetable or animal. Some are shown in Figure 13.2. The vegetable fibers are mainly composed of cellulose. The vegetable fibers are prominently employed for application of *composites*. The animal fibers are mainly composed of proteins (Verma et al., 2016). The natural fibers have a cell structure consisting of tangled fiber connected to rigid cellulose microfibrils and surrounded in a hemicellulose matrix. The vegetable fibers are harvested from the stems, leaves or various types of plant seeds, hence classified as a distinctly renewable resource.

13.2.2 BIOPOLYMERS

Biodegradable biopolymers are derived from renewable sources and they may include the thermosetting and thermoplastic polymers. Since these materials are made by one repeating unit of monomer which may be synthetic, thermosets are formed by polymerization, hence they aren't 100% green materials. Also starch derived thermoplastic is another polysaccharide commonly used as a matrix for green composites. Proteins such as albumin and collagen are also biopolymers. The classification of natural and bio-fibers is shown in Figure 13.3.

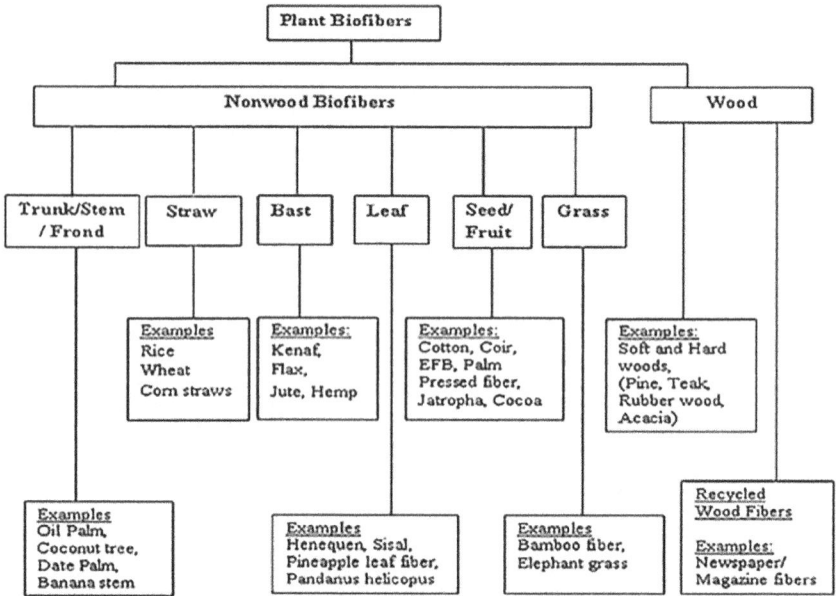

FIGURE 13.2   Different plant based natural green composites and their sources.

Source: Abdul Khalil et al. (2012)

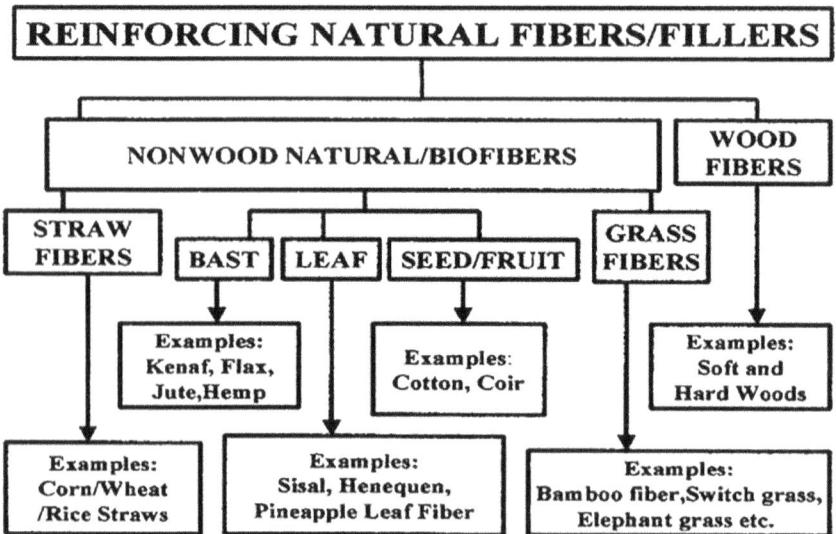

FIGURE 13.3   Classification of natural/bio-fibers.

Source: Bharath & Basavarajappa (2016)

13.2.3   CLASSIFICATION OF NATURAL OR BIO-FIBERS

13.2.4   WHY TO USE GREEN COMPOSITES

Biodegradable materials are sustainable products. Studies have found that by 2009 around 90% of the plastic made product available in the market were made from non-renewable resources and they are mostly non-biodegradable, filling up the land-fills at an alarming rate. Their disposal not only increases the costs, but also availability of space is limited. Till now there have been many management strategies that have been implemented which include principles such as recycling, reusing or reducing or sorting, mechanical recycling, chemically recycling and incineration. Also disposing of plastics generates a huge amount of carbon dioxide and other greenhouse gases. It is estimated that 5.1 tonnes of carbon dioxide is generated by plastic wastes from landfills. (*Biodegradable Green Composites—Google Books*, n.d.).

All these issues make it very important to either make plastic safe with proper-ties such as enhanced lifespan, reuse them, find ways to dispose of them safely or replace them with other alternatives that would be biodegradable in order to protect the non-renewable resources and hence the ecological balance (Bismarck et al., 2006). The biggest problem of plastic is its waste management which is generated during its lifetime. To enhance and to make better materials, alternative composites should be used which may have environmental protection as well as be eco-friendly, with low or similar cost of production. Eco-friendly designs of products should be made, consider-ing their end life scenarios. For sustainable development recycling, reuse and remanu-facturing should be considered for finding alternatives. Similarly, good materials that are environment friendly and also can optimize the use of energy should be used, and this would eventually reduce the landfill spaces (Dicker et al., 2014b).

Green composites provide the require properties, which are needed to be replaced. They are biodegradable and are mostly produced from renewable resources waste products. They are environment friendly and can be disposed of very easily. To further enhance the performance, in natural fibers the product development and its operational life must have the possibility of reducing the weight and also avoiding the material dis-posal in landfills; their application is being verified in cotton fibers which are used in vehicular application. Some of the recycling methods are given in Figure 13.4. These features may help to reduce energy consumption with the product's durability. Also as a result, it expands and enhances the life cycle of the product, postponement of dump-ing of wastes and side-steps emissions. Green materials create the pathway for future products. Sustainability requires exploration of new products and unique methods that are environment friendly.

13.3   CARBON FOOTPRINTS

The term "carbon footprints" has been frequently used in previous years. With the focus on recent climate changes, carbon footprints is commonly used across media, government and the business world. Carbon footprints is a widely used term in regards to global climate change. "Carbon footprints" can be used to describe the portion of gaseous emission that is crucial to climate change which can be linked with man-made

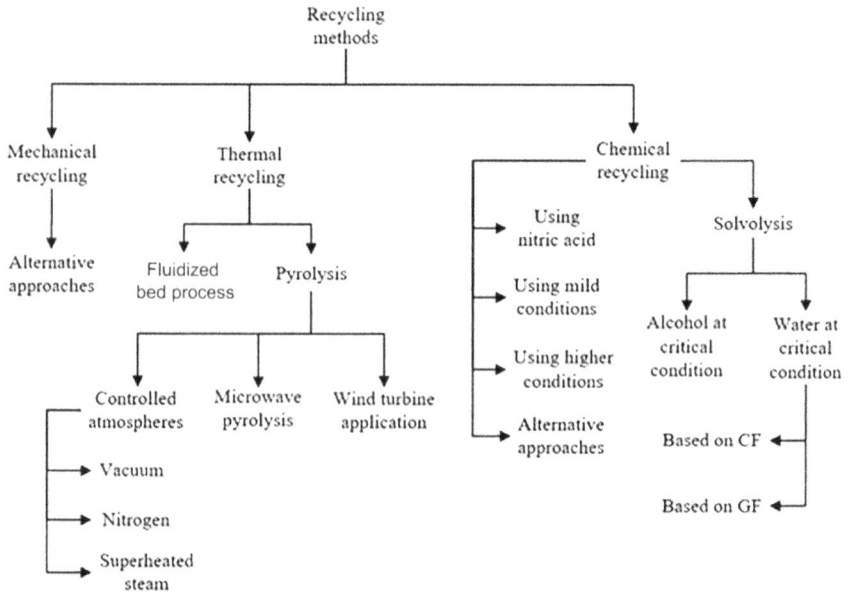

FIGURE 13.4   Different recycling methods that are used for green composites.

Source: Karuppannan Gopalraj & Kärki (2020)

activities. The spectrum of definitions can range from the direct emission of carbon dioxide gas to a full life-cycle greenhouse gas emission (Pertsova, 2007).

The term "footprint" suggests a measurement or expression in area-based units. Closely related to carbon footprints are "ecological footprints" which are measured in hectares or "global hectares." This is an interesting proposition to decide if carbon footprint should be a pressure indicator which would express the amount of carbon emission or if it would indicate the impact and quantify the amount of carbon dioxide released. In some cases carbon footprints can be used as a broad substitute for the emission of greenhouse gases which includes carbon dioxide. Some researchers describe carbon footprint as the "carbon weight" which is produced per person or an activity in kilograms and tonnes (Lu et al., 2019). The carbon footprints may hence be defined as a degree of the quantity of carbon or carbon dioxide emission which is caused directly or indirectly by an activity or an individual, which is accrued over the life stages of a product.

In general the activity of an individual, population, company, industry, government, organization etc. produces or uses good and services that can have direct or indirect carbon emission can be taken into account. A greenhouse gas indicator should include all the types of gases that may include carbon dioxide or methane which are relevent and most comman greenhouse gases and also other substance which have global warming potential. But due to less data availablity, it is difficult to quantify them. Therefore in these cases we can only include carbon dioxide as the most practical and clear greenhouse gas taken into account to have a carbon footprint (Pertsova, 2007).

13.4   LIFE CYCLE ASSESSMENT

The life cycle assessment (LCA) is basically a tool feature which can be employed to determine the total environmental impact of a product throughout the period of its lifetime. The life cycle assessment, that is, LCA helps in comprehending the various available material, their proper selection which would help the manufacturer to select a given material and also finding its replacement which would then improve the environmental aid of the material. The main concern about the life cycle assessment is the impact of the greenhouse gases, that is greenhouse gasses which mostly consist of CO_2. Due to the global concern about environment protection, an indicator is used to measure the impact of the carbon dioxide emissions produced internationally by different companies. The indicator is used to measure the carbon footprints (Vilaplana et al., 2010). However, there are certain limitation of such indicators that must be documented for specific cases. The limitation arises due to the standard and diverse methods which are used to study the life cycle (Correa et al., 2019).

The life cycle assessment (LCA) is basically a tool feature that can be used to determine the total environmental impact of a product throughout its lifetime. The life cycle assessment is an essential tool, which can be used to design new products. To produce sustainable materials life cycle assessment is the most important tool, as without considering its impact through its lifetime it would not be possible to produce a sustainable eco-friendly material (Correa et al., 2019). It is used to estimate the inputs and outputs that are needed of the resources for a specific process or product. The LCA can be used to study the environmental impact of any product as compared to other similar products. The different stages of LCA are used to assess, determine and improve the composites for a sustainable approach (Ead et al., 2021). The LCA technique permits determination of a comprehensive preview of all the environmental perspectives relating to manufacturing techniques and influences the mass and energy flows as well as its life cycle. All these steps and flows are then linked to their direct and indirect environmental impacts, which must be determined quantitatively. The details are given in Figure 13.5.

13.4.1   RAW MATERIALS

To study the life cycle, the first step is to recognize the information about the raw materials. The acquisition regarding the raw material is the first stage. For the natural fiber bio composite materials the main materials which are used are polymers as matrix and natural fibers as fillers. The natural fibers are generally obtained from the residue of agronomic production. The farming-agriculture processes include sowing, harvesting and fertilization; fertilizers are the biggest contributor to the emissions of greenhouse gases. Depending on which type of fertilizers used, the emission of such greenhouse gases for the crop can be calculated and it also depends on the weather conditions in the region surrounding the crop (Correa et al., 2019).

Depending on the interaction of the different interfaces such as atmosphere, soil and plants the different dynamics of carbon dioxide emission from the plants can be calculated. Plants release carbon dioxide via the process of photosynthesis and the soil contains different carbon components in its organic forms. A negative contribution is

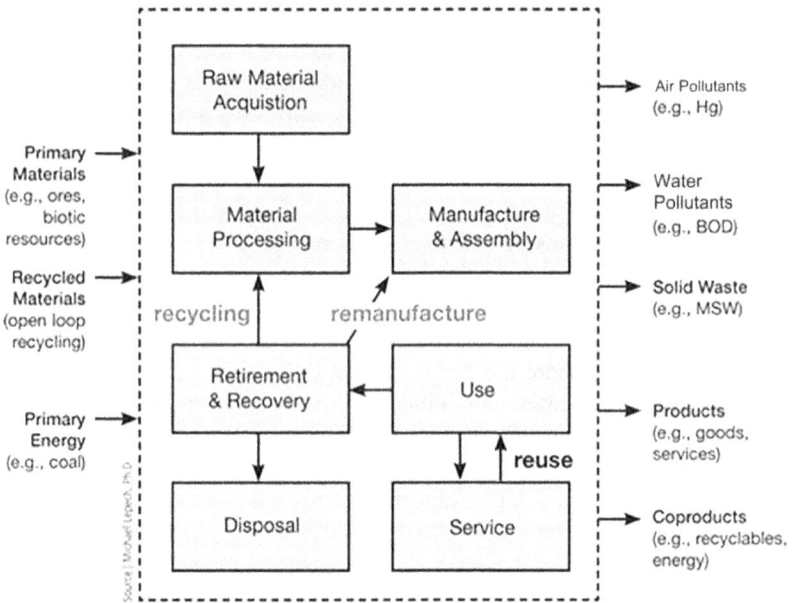

FIGURE 13.5 Lifetime assessment of green composites having different stages.

Source: Life Cycle Assessment: Are Composites "Green"? I CompositesWorld (n.d.)

observed due to the type of the vegetation, the soil and also the climatic condition. The carbon balance that is the balance between the carbon released to the carbon reinstate by the plants is also affected and the release of carbon is found at a higher rate than the later part. An imbalance can be caused by this distinctiveness because of the embodied carbon throughout the process of farming-cultivation and also the resulting emission during the fiber production, thus stopping the evaded emissions form carbon reserves (Correa et al., 2019).

Looking at the process of polymer production, various natural resources such as water, energy and chemicals are used in very large amounts during the stage of manufacturing. The huge number of wastes produced causes a very significant impact on the environment and also on human health. There are various techniques that are used to manufacture the polymers, and each process may be considered as part of the manufacture line that marks the incidental conditions of the material. The raw material should be used after studying the emission factor which can be evaluated by studying the life cycle of given composites. Different approaches can be considered for analyzing the average emissions for different types of polymer materials which contribute greatly to the all-inclusive analysis for the emissions of base or raw material. We are analyzing the biological-based material systems, the matrix which is used is mostly biodegradable, that is it detoxicates on its own naturally and it should be considered as an eco-friendly alternative, keeping in mind the environmental sustainability and ecological impact.

13.5 CARBON FOOTPRINT INDICATOR

The carbon footprint indicator belongs to a group of indicators which is designed to assess the environmental stress which is caused by the anthropogenic activities. This indicator allows to particularly evaluate and quantify the production and services in the amount of carbon dioxide released to the atmosphere (Correa et al., 2019). While counting the carbon footprint, one certainty that arises is double counting and another is the type of inputs that are allowed. The double counting of emissions all through study of a case can be there around the indicator. The drawback in calculation of carbon footprints via this method could be the calculation mistake that may get raised due to the involvement of the different life cycle methods that are used in the supply chain for the production of an explicit entity, which is used in production of composite materials, as the feedback process to originate different articles is used which may lead to double calculation of the carbon footprints. Among the various studies which are conducted, one of which is in between the years 2006 and 2007; it was concluded for quantification of carbon footprint. This may include the methodological procedure to monitor the objectives, operational and organizational limitation and also carbon footprint quantification (Correa et al., 2019).

These bio-based products are mostly biogenic in origin which has carbon embodied in it. Hence, quantifying carbon can be done considering two main approaches which are that the biogenic carbon can be omitted from the investigation while keeping in view of carbon dioxide emissions unbiased; or the biogenic carbon can be well-thought-out carbon storage vis-à-vis the carbon dioxide emissions reserved during photosynthesis processes (Correa et al., 2019). Application of such indicators can be done by taking into account the understanding of carbon dynamics below the artefact.

The composite of natural fibers is found to contribute to the lower emission. This is because of the fact that they have a substantial potential in the direction to absorb carbon dioxide which hence helps in providing good storage facility for the emanations and influence to lower the carbon dioxide production and its concentration in the environment.

Hence to completely understand and represent the assessment of analyzing the carbon footprints of any product under observation, all of its stages of life are equally important to be studied for this analysis. This process would further help in defining and considering all the various possible activities that can lead to a contribution to the production of carbon emission to our atmosphere and thus contribution to carbon footprints in environment.

13.6 CARBON FOOTPRINT ANALYSIS

The analysis of carbon footprint can majorly be divided into four phases which are the goals, lifetime inventory, lifetime impact evaluation and the interpretation of life cycle. Following these standard phases, the study for carbon footprint assessment can be thus divided into four stages which are: the composite characterization, goals and future scope, the life cycle assessment of carbon dioxide and lastly the life cycle impact assessment (Caldas et al., 2017) as shown in Figure 13.6.

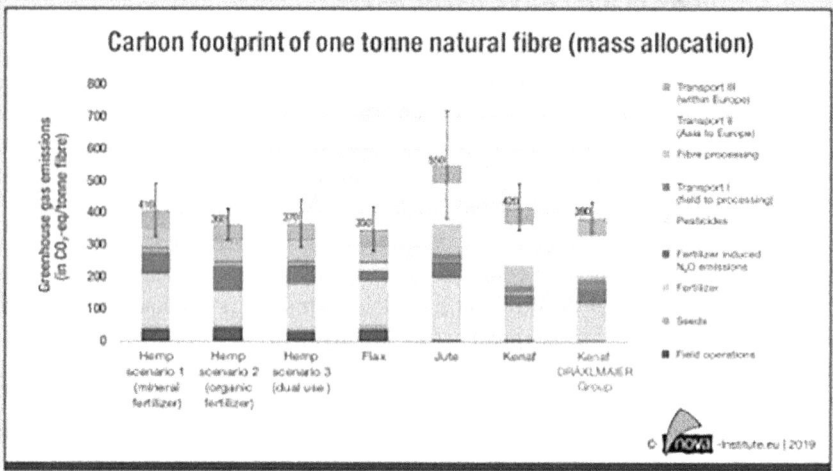

FIGURE 13.6 Graphical data showing carbon footprint for different natural fibers.

Source: Renewable Carbon News (n.d.)

13.6.1 COMPOSITE CHARACTERIZATION

Herein composites can be prepared from plant based raw materials which may include bamboo, rice, husk or wood shaving. These composites are treated at NUMATS–center of sustainable materials and federal university of Rio de Janeiro, Brazil as they are mostly considered as waste materials of the industrial process. The water absorption property of these material was used to synthesize uniform composite. As an example, the composite of rice husk has been found to absorb maximum water.

13.6.2 GOALS AND SCOPE

The aim is to analyze the carbon footprints while considering the carbon emissions and removals (Korol et al., 2020). The carbon footprint boundary provides a support for gate assessment since it is restricted to synthesis process in composite. Along with biogenic carbon emissions, the end-of-life stage is also taken in consideration (Thakur et al., 2018). In the life cycle of biodegradable materials, specially forest-based materials, the knowledge about carbon footprint is a critical issue, particularly the emissions from biogenic carbon products. Storage capacity as well as delayed carbon emission used in quantification and consideration in the end-of-life (Dey & Ray, 2018).

The carbon stored by the process of photosynthesis is re-emitted into the atmosphere, totally or partially considering the life cycle. In case of landfilling, a huge amount of biogenic carbon is captured in ground for a long time without degradation (or very low degradation). To overcome this issue, carbon storage is an appropriate method (Kumari et al., 2007).

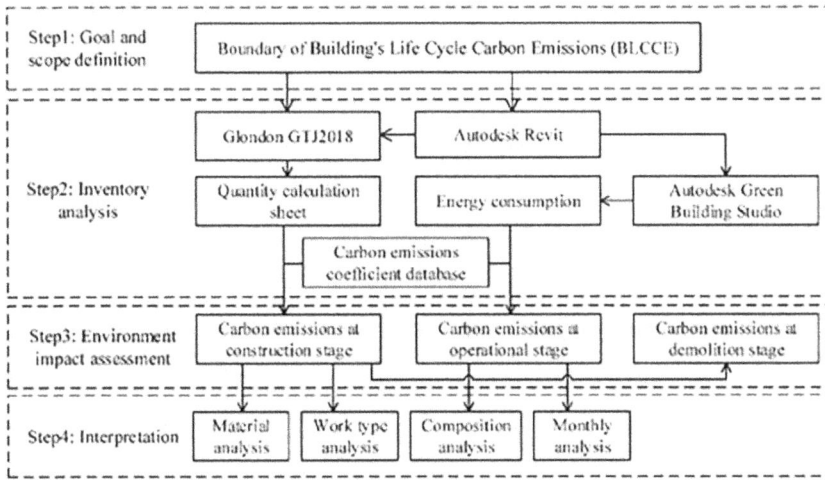

FIGURE 13.7   Carbon footprint analysis framework.

Source: Lu et al. (2019)

13.6.3   LIFE CYCLE ANALYSIS CONSIDERING CARBON DIOXIDE INVENTORY

The carbon dioxide from the biomass-based materials can be used in the cultivation process instead of considering it as waste product.

The content of biogenic based carbon (M_{CO2}) can be computed as:

$$M_{CO2} = m_{dry} \times C \times \frac{mm_{CO2}}{mm_C}$$

Here M_{CO2} = mass of CO_2 sequestered (kg)—biogenic carbon; M_{dry} = dry mass of the bio-based material (kg); C = percentage of carbon in dry matter (%); mm_{CO2} = molecular mass of CO_2; mm_C = molecular mass of carbon. The percentage of carbon in dry matter values by elemental analysis of CHN (determination of carbon, hydrogen and nitrogen) helps in calculating the biogenic carbon mass (Korol et al., 2020).

These green composites are not exclusively eco-friendly compatible because of issues like recyclability at higher temperatures and biodegradability in respect of fillers (Mohanty et al., 2002). Various polymers like polysaccharides, polyesters, proteins and polycaprolactone exist which are capable of biodegrading (Vilaplana et al., 2010). Figure 13.7 shows carbon footprint analysis framework having different steps.

13.6.4   LOW CARBON BIO-BASED MATERIALS SPECIFICATION

It has been highlighted by various studies that the important parameters for a small carbon footprint bio-based material design can be as follows:

1) Lowering cement or matrix material concentration
2) Increasing proportionately biomass concentration

3) Selection of appropriate bio-based material with higher carbon content
4) Selection of bio-based material with low CO_2 emissions production
5) Selection of bio-based material having workability at the fresh state
6) Selection of bio-based material available close to manufacturing unit

13.7 ADVANTAGE OF CARBON FOOTPRINTS

1) Analysis of carbon footprints enables an inclusive and complete evaluation of greenhouse gases emission.
2) It obeys given environmental and economic reporting standards.
3) It comprises all gases possessing greenhouse potential.
4) The emission figures are analogous and accessible to other regions.

13.8 DRAWBACK OF CARBON FOOTPRINTS

1) It caters to global warming alone and neglects equally other crucial environmental facets.
2) It is unable to track the complete spectra of human needs in respect of the environment.
3) The supplementary evaluation techniques are required to assess the effect of climate change at national and subnational levels.

13.9 CONCLUSIONS

The research on bio-composites with new development is constantly rising and is in demand. The green composite products provide renewability and also have great potential in reducing the greenhouse gasses emission. The green materials are excellent contender material for making sustainable products. Considering the present-day scenario, producing a sustainable product is a very complex concept which may involve various considerations and assumptions with one objective that is to achieve and explicitly intricate various factors needed to produce sustainable composite, which are not suitably demarcated otherwise, considering the life cycle assessment and environmental performances of products. The manufacturing techniques, raw materials assortment and the transportation are a few factors which are some of the adjustable features according to which the product or material is being developed (Correa et al., 2019).

The natural fibers can be used with other composite matrixes, which would establish many advantages such as lowering the emission of greenhouse gases to several diverse products. This reduction in the production of gases helps in evocative enhancement due to the negative influence from different activities such as the consumption of energy, dispensation and discarding procedures thus reducing the carbon footprints (Caldas et al., 2017).

For future use the green-composite products can be essentially considered as a sustainable material in contrast to the various conventionally used materials which are based on non-renewable resources such as fossil fuels or polymeric based materials. With this, there can be few characteristics; for example, the process used during the amplification of the materials during its life cycle can lead to different types of environmental impacts, which further can be considered as inferior to the ones momentous from old-fashioned composite elaboration.

REFERENCES

Abdul Khalil, H. P. S., Bhat, A. H., & Ireana Yusra, A. F. (2012). Green composites from sustainable cellulose nanofibrils: A review. *Carbohydrate Polymers*, *87*(2), 963–979. https://doi.org/10.1016/J.CARBPOL.2011.08.078

Bharath, K. N., & Basavarajappa, S. (2016). Applications of biocomposite materials based on natural fibers from renewable resources: A review. *Science and Engineering of Composite Materials*, *23*(2), 123–133. https://doi.org/10.1515/SECM-2014-0088/MACHINE-READABLECITATION/RIS

Biodegradable Green Composites—Google Books. (n.d.). https://books.google.co.in/books?hl=en&lr=&id=WsuICwAAQBAJ&oi=fnd&pg=PA312&dq=green+composite+materials&ots=5vRMmQp3r3&sig=kMr08VdHMukXr51DFLdUoyEU4Q8#v=onepage&q=green%20composite%20materials&f=false

Bismarck, A., Baltazar-Y-Jimenez, A., & Sarikakis, K. (2006). Green composites as panacea? Socio-economic aspects of green materials. *Environment, Development and Sustainability*, *8*(3), 445–463. https://doi.org/10.1007/S10668-005-8506-5

Caldas, L., da Gloria, M., Santos, D., Andreola, V., Pepe, M., & Toledo Filho, R. (2017). Carbon footprint of bamboo particles, rice husk and wood shavings-cement composites. *Academic Journal of Civil Engineering*, *35*(2), 499–506.

Correa, J. P., Montalvo-Navarrete, J. M., & Hidalgo-Salazar, M. A. (2019). Carbon footprint considerations for biocomposite materials for sustainable products: A review. *Journal of Cleaner Production*, *208*, 785–794. https://doi.org/10.1016/j.jclepro.2018.10.099

Dey, P., & Ray, S. (2018). An overview of the recent trends in manufacturing of green composites-considerations and challenges nomenclature MSW municipal solid waste WPC wood plastic composite MPa mega-pascal GPa giga-pascal. In *Pritam Dey and Srimanta Ray/ Materials Today: Proceedings* (vol. 5). www.sciencedirect.comwww.materialstoday.com/proceedings2214-7853

Dicker, M. P. M., Duckworth, P. F., Baker, A. B., Francois, G., Hazzard, M. K., & Weaver, P. M. (2014b). Green composites: A review of material attributes and complementary applications. *Composites Part A: Applied Science and Manufacturing*, *56*, 280–289. https://doi.org/10.1016/J.COMPOSITESA.2013.10.014

Ead, A. S., Appel, R., Alex, N., Ayranci, C., & Carey, J. P. (2021). Life cycle analysis for green composites: A review of literature including considerations for local and global agricultural use. *Journal of Engineered Fibers and Fabrics*, *16*. https://doi.org/10.1177/15589250211026940

Pertsova, C. C. (2007). *Ecological economics research trends*. Nova Science Publishers, New York.

Karuppannan Gopalraj, S., & Kärki, T. (2020). A review on the recycling of waste carbon fibre/glass fibre-reinforced composites: Fibre recovery, properties and life-cycle analysis. *SN Applied Sciences*, *2*(3), 1–21. https://doi.org/10.1007/S42452-020-2195-4/TABLES/5

Khalid, M. Y., al Rashid, A., Arif, Z. U., Ahmed, W., Arshad, H., & Zaidi, A. A. (2021). Natural fiber reinforced composites: Sustainable materials for emerging applications. *Results in Engineering*, *11*, 100263. https://doi.org/10.1016/J.RINENG.2021.100263

Korol, J., Hejna, A., Burchart-Korol, D., & Wachowicz, J. (2020). Comparative analysis of carbon, ecological, and water footprints of polypropylene-based composites filled with cotton, jute and kenaf fibers. *Materials*, *13*(16). https://doi.org/10.3390/MA13163541

Koronis, G., Silva, A., & Fontul, M. (2013). Green composites: A review of adequate materials for automotive applications. *Composites Part B: Engineering*, *44*(1), 120–127. https://doi.org/10.1016/j.compositesb.2012.07.004

Kumari, R., Ito, H., Takatani, M., Uchiyama, M., & Okamoto, T. (2007). Fundamental studies on wood/cellulose-plastic composites: Effects of composition and cellulose dimension on the properties of cellulose/PP composite. *Journal of Wood Science*, *53*(6), 470–480. https://doi.org/10.1007/S10086-007-0889-5/METRICS

La Mantia, F. P., & Morreale, M. (2011). Green composites: A brief review. *Composites Part A: Applied Science and Manufacturing*, *42*(6), 579–588. https://doi.org/10.1016/j.compositesa.2011.01.017

Life Cycle Assessment: Are composites "Green"? I Composites World. (n.d.). www.compositesworld.com/articles/life-cycle-assessment-are-composites-green

Lu, K., Jiang, X., Tam, V. W. Y., Li, M., Wang, H., Xia, B., & Chen, Q. (2019). Development of a carbon emissions analysis framework using building information modeling and life cycle assessment for the construction of hospital projects. *Sustainability*, *11*(22), 6274. https://doi.org/10.3390/SU11226274

Mann, G. S., Singh, L. P., Kumar, P., & Singh, S. (2018). Green composites: A review of processing technologies and recent applications. *Journal of Thermoplastic Composite Materials*, *33*(8), 1145–1171. https://doi.org/10.1177/0892705718816354

Mohanty, A. K., Misra, M., & Drzal, L. T. (2002). Sustainable bio-composites from renewable resources: Opportunities and challenges in the green materials world. *Journal of Polymers and the Environment*, *10*(2).

Thakur, S., Chaudhary, J., Sharma, B., Verma, A., Tamulevicius, S., & Thakur, V. K. (2018). Sustainability of bioplastics: Opportunities and challenges. *Current Opinion in Green and Sustainable Chemistry*, *13*, 68–75. https://doi.org/10.1016/j.cogsc.2018.04.013

Verma, D., Gope, P. C., Zhang, X., Jain, S., & Dabral, R. (2016). Green composites and their properties: A brief introduction. *Green Approaches to Biocomposite Materials Science and Engineering*, 148–164. https://doi.org/10.4018/978-1-5225-0424-5.CH007

Vilaplana, F., Strömberg, E., & Karlsson, S. (2010). Environmental and resource aspects of sustainable biocomposites. *Polymer Degradation and Stability*, *95*(11), 2147–2161. https://doi.org/10.1016/j.polymdegradstab.2010.07.016

Zini, E., & Scandola, M. (2011). Green composites: An overview. *Polymer Composites*, *32*(12), 1905–1915. https://doi.org/10.1002/pc.21224

14 Life Cycle Assessment of Eco-Friendly Composites

Shivani A. Kumar, Gaurav Sharma, Tanika Gupta, and Rohit Verma

CONTENTS

14.1 INTRODUCTION

With the ever-increasing escalation in human-induced changes in weather, there is a desire for initiatives taken to reduce the effect of greenhouse gases (GHG) on our environment. One of our steps to reduce the value of carbon dioxide and various greenhouse gases (GHG) emitted is to stabilize commercial production systems and to take into account the environmental impact of these products on the production, applications, and detoxification. Although industrialization has allowed for less mass production, the large number of raw materials utilized and the waste

DOI: 10.1201/9781003272625-14

product produced contribute to reduction of many products and their impact on environment.

Life Cycle Assessment which may be abbreviated as LCA is understood as a tool which may be implemented to decide the total environmental effect of the product at some point of its lifetime (Gupta & Kumar, 2010). LCA began in the '60s, but it originated as an ambitious scientific process in the 1990s, due to the need for big players in the food and beverage industry to gain a more detailed understanding of the implications of their product packaging. Concerns have been raised about the use of LCA information in marketing rather than as a decision-making tool. This led to the development of the international standard ISO 14040, which regulates LCA practice and lays the foundation for what is considered to be a good assessment and somewhat useless. A very important factor inside LCA is monitoring and calculating the insource and outsource for some particular product or their method. When LCA is finished, the environmental impact of the product may be in comparison to a few different products and it may be improved as well. The precise levels of a material's life cycle may be computed to decide wherein maximum resources are utilized or the maximum GHG are required. This helps in bringing the approach to be stepped forward upon closer to a maximum sustainable method.

Fiber-reinforced products are materials consisting of fiber bunch enclosed in a matrix cloth. In the evaluation of these composites, we are capable of recognizing the polymeric fiber reinforced composites in which, normally, a polymeric resin is used as the matrix material. Polymeric resins are homogeneous and isotropic which proves to be very beneficial in polymeric fiber (Curran, 2016). Composite substances are manufactured and produced to collect the houses in their substances to supply advanced stop-product. Artificial fibers are standardly applied in those products as their residences are more constant and tractable than other basic fibers. However, artificial products can also have a massive, terrible environmental effect (Thankachen et al., 2021). Composites made from artificial fibers are neither biodegradable nor recyclable at the stop in their life cycles due to the distinguishability between the fibers and the matrix. To extend with, maximum composite materials are artificial polymer matrices making use of petroleum. These artificial polymers which are also non-biodegradable matrices contribute to global warming, enhance the landfill deposits, and give rise to poisonous environmental consequences. Table 14.1 shows the cultivation regions and cultivation conditions for a few common natural fibers.

In the evaluation, herbal fibers degrade when given their presence or can be used to restore energy. Furthermore, if the matrix is also biodegradable, the whole mixture may decompose concluding its useful life. The use of biopolymer matrices derived from renewable sources of products additionally reduces vulnerability of humans on some renewable sources of fossil fuels (Tukker, 2000). The bio-compound is sourced from natural basic fiber reinforcement or an herbal matrix "seasoned" with blended herbal fiber and natural matrix. Therefore, it is promising to develop "innocent" materials which remarkably reduces environmental impact as compared to synthetic-based finishing composites. LCA is also applicable on current environmental impact between herbal fiber blends and synthetic fiber blends at all levels of the survival point. However, the difference is basically very crucial in determining stability of innocent compounds that need to be further developed.

Currently, there is limited evaluation study in this field for the placement of herbal fibers, bio-composites, and innocuous compounds. In addition, current LCA reviews

TABLE 14.1

Cultivation Regions and Cultivation Conditions for a Few Common Natural Fibers

Fiber type	Cultivation regions	Cultivation conditions
ABACA	The abaca trade and production is led by the Philippines with 60% market share, followed by Ecuador with 35% market share.	It usually grows in tropical areas. Fine growth of fibers occurs at 28°C–30°C with good relative humidity and 2000–2500 mm of water per year. The first fiber takes about 2 years to produce, and the next harvest takes place every 2–3 months.
Bamboo	Almost all bamboo production occurs throughout the Asia-Pacific region. The major producers of bamboo are China, India, and Brazil.	It grows among many different climates, from forests to mountains. It is well-grounded in moist soil with good water holding capacity. Snow should be avoided, and annual rainfall should be no more than 1000 mm.
Flax	Popular agricultural areas include France, Canada, Argentina, India, USA, Ukraine, Kazakhstan, Egypt, and the Czech Republic.	High humidity relative to temperatures between 18°C and 20°C. Good soil conditions include fertile soil with a flexible compact structure for air access and a pH of 6.5–6.9.
Hemp	It is common to grow in various regions around the world, including cool, tropical, and subtropical areas. Most hemp production is found within Canada, China, and the European Union (EU). France, the Netherlands, and Romania are the largest producers in the EU.	It grows between 5.6°C and 27.5°C, and 14°C is ready for growth. It takes 4–5 months without frost to produce harvested crops. Good yields require 500–700 mm of rainfall. The ideal soil pH is raised to 6.0.
Jute	Equatorial, tropical, and subtropical zones. Grown in India, Bangladesh, China, Vietnam, Myanmar, Thailand, and Brazil.	Hot and humid weather during wet summers; 24°C–27°C is ideal for annual; rainfall of 1000–2000 mm of rainfall at sowing time.
Kenaf	Typical production regions include India, China, and Pakistan.	Production fits well worldwide with high environmental flexibility. It needs a soil temperature of about 15°C to grow.
Ramie	Most production takes place in China. It is grown in India, Indonesia, Korea, Taiwan, Brazil, Japan, and the Philippines.	A humid climate with moderate temperatures. The relative humidity of 25% and temperatures of 20°C–31°C is ideal. It needs equally divided rainfall of 1500–3000 mm per year.
Sisal	It is thought to be from Central America. Most production occurs in Brazil and Tanzania, although China and Kenya are major producers.	It grows in tropical or subtropical climates on well-drained soil anywhere between sea level and snow line. It requires high humidity with high temperatures. Sisal consumes lime, magnesia, potash, and phosphoric acid.

are restricted to selected polymers/bio-composites, although not innocuous com-pounds. For this reason, this evaluation provides and examines the demands of LCAs when posted to linear composites, however, at the same time including factors such as economic impact (survival cycle cost) and toughness. The absence of study on LCAs is an obstacle to providing innocuous mixture as an important alternative to synthesis of compounds. In addition, this evaluation gives an idea of choosing the form of basic fiber taking into account the area. It is being considered for an investi-gation into natural fiber production in Argentina and Canada.

14.2 THEORY OF LIFE CYCLE ASSESSMENT

The LCA (Netravali et al., 2003) is "a logical machine specially manufactured to exam-ine the environmental effect of an object's overall production chain." To certify an overall analysis, the LCA also studies the effects of the materials during utilization and through-out life. We can say, LCA is a very broad machine which has a wide application on a wide variety of industries, sectors, and product types. Therefore, the guideline does not specify how each person will complete the step because the correct procedure varies widely between the products. Instead, the company meant for standardization main-tained a simple rule to deliver the stages required to reach the LCA (La Rosa et al., 2013).

The LCA will help determine if the newly made composite product is more or less environmentally friendly than the previous one. In production of LCA, the life stages of the product are divided so that the data are calculated for each important origin and output of all expedient and energy. This process is crucial to the manufacturing procedure which produces in the stages of the material life cycle in which environmen-tal impacts are likely to improve (Rebitzer et al., 2004). Accordingly, LCA is a very crucial stage taken in the origin of many green alloys because it allows the analysis of materials with far fewer environmental effects than their synthetic counterparts.

14.3 BASIC REQUIREMENTS

According to ISO, the basic study process for LCA framework (ISO, 14040, 2006a; ISO, 14044, 2006b) is designed as shown in Figure 14.1.

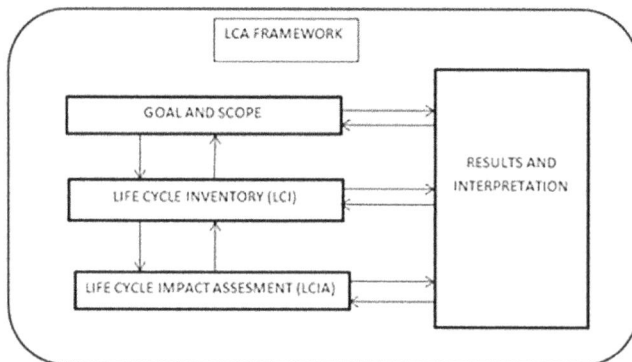

FIGURE 14.1 The basic study process for LCA framework (ISO 14040, 2006; ISO 14044, 2006).

(1) Goal and scope,
(2) Lifecycle Inventory (LCI),
(3) Lifecycle Impact Assessment (LCIA), and
(4) Interpretation of results.

The aforementioned steps are discussed in detail next.

14.3.1 STEP I: GOAL AND SCOPE

According to the International Organization of Standardization, the initial step for the LCA is to define and understand the research goal. This always deals with the quality of analysis and also application of its results. Our study should also be noticed, which must include setting system parameters. A system parameter is defined as the system which is excluded out of environment to track all the in stage and out phase. Within the boundaries of the system, the analysis of composite life cycle must be mentioned and considered. Many studies use the method defining procedure from harvesting to dump (Mansor et al., 2015). However, it is a very crucial step to separate the harvesting margin and end-of-life stages involving jungle, grounds, and landfills. Consistency is also important while finding the extraction of the system.

A working unit is set in this stage. An active system is defined as a fixed quantity of composite to be used and read. By setting up an operating unit, all inputs and outputs of all procedures in the system can be distinguished and integrated. In addition, the operating unit is very beneficial in the comparative studies of two separate factors (Mansor et al., 2018). Here, the total in stage and out phase values are equal between these two types of objects to facilitate comparisons of individual environmental impacts. Although the operating unit is usually the product value, it can be the time limit for the product to be used. This depends on the performance of material. Pegoretti and his coworkers (Pegoretti et al., 2014) have shown time-based performance which compares many acoustic systems in the Canadian automobile sector. One of them produced synthetic plastic while another was derived from recycled fiber whose purpose was explained as to "examine" and to "comparison" the important impact of three acoustic panels. According to the study estimates, the LCA would look at these three stages: (1) production, (2) use, and (3) end-of-life. They all will be able to maintain an acceptable optimum level in their use for 12 years.

14.3.2 STEP II: LIFE CYCLE INVENTORY (LCI)

The next phase of the LCA is the analysis of Life Cycle Inventory (LCI). LCI is required to identify and calculate resources for the unit. The aim of LCI is to compile a list of all the inputs and also the outputs of materials, their waste, and original source for every procedure with respect to the working unit (Ramesh et al., 2020). Unit processes, which are the basic stages in the life of a product can be combined to form a production system. They may also have single or group operations. In the next step, all water flow, energy, and raw resources are discharged into the atmosphere, land, and water. It is also important that system boundaries are detailed and consistent. A chart is then created to visualize and link flows for these processes. A current is calculated to relate the functional unit with data accuracy and stability. However, due to the large numbers, it is not practical for the LCA to have a complete database. Therefore, databases can be very common for the whole analysis or the LCA. The data can be collected. After the data during the process and it is maintained as an inventory for further analysis.

PS/PET and PLA are classified as petroleum-based and bio-based plastics respectively. In the current study, the system is divided into a few unit processes, which may be compiled into two common stages in the life cycle of a product: one is production and the other is waste management. Now, data are collected for boxes from a previous assessment study. These boxes are used for waste management which is divided into sanitary landfills and controlled composting. Shipping cost is also calculated depending on the miles travelled and the quantity of composite. Garbage sanitization landfills waste and composts are received from a preliminary study. Both LCI and LCA involve a direct financial cost and should not be compared with any other life cycle approach. An industry called "I/O-LCA" is a method based on input/output and concentrates on combining monetary assistance with environmental evaluation.

14.3.3   Step III: Life Cycle Impact Assessment (LCIA)

The next stage in the LCA process is the Life Cycle Impact Assessment (LCIA). The main motive of the LCIA is searching and calculating the total environmental impacts of the process and product. This also depends on the final calculations in the LCI, as the result of the LCI is lengthy and entangled with new calculations. For this, the LCIA changes inventory analysis results into common units within a wide range of effects. These are categorized into different steps in which the initial steps are important. These are (1) impact categories, (2) impact category indicators, (3) division of list results into sub parts, (4) characterization, (5) generalization, (6) accumulation, and (7) loading.

First, select the effect categories. Impact categories are collections of data of environmental factors that spoil category. Influence steps should also relate the scope and its study. Companies include three main divisions of impact: human health, ecosystem health, and resource damage. When LCA undergoes with these categories, the study is called an "end-point damage" model. Model name "Mid-Point Damage" is applied to design impact categories in a significant manner. For example, many of them are utilized as climatic control, ozone damage layer, human poisoning, and acidification. Furthermore, land use is a more frequently used category of influence. While this may seem unnecessary, land use can have a significant impact to be considered because the used area is a restricted source. After the establishment of impact, they are categorized as models, category indicators, and other characterization factors. This definition depends upon globally proved model of local segment issue. The third step involves the process of allocation of inventory results. It helps in identifying impact categories. This is distinct in an implicit category in a better manner for future survey. In the fourth stage, division takes place at all factors for each effect category. These are presented as a simple unit, to integrate numbers and to get the final calculated value of the category. This is termed as category index result.

The first four steps in the LCIA are to prepare information to be identified if necessary, and three other steps can be taken in relation to current tradition and their past analysis. Generalization, which is the first optional stage, involves expressing range index results relative to contextual data. For example, a heavy range can cause less

serious spoilage comparative to small value range. It may be due to the effect categories with dissimilar units. This may result in different values of characterization rather
than the actual effect. Later group stage, the accumulated values of impact groups are
sequenced by importance ranking of their impact on the final step. Every result implies
generation of a numerical waiting factor dependent on the assigned important stage.
This result when exaggerated by the index result gives the weighted value.

14.3.4   Step IV: Interpretation of Results

The last stage of the LCA is the total prediction of the results in the LCI and LCIA
on the basis of objective. This phase determines the calculated value and yields the
results. Then the conclusions can be made and recommendations are interpreted.
Determining the stability and reliability of data and process is part of this analysis. The
limitations provided of the theories should also be discussed. In order to fulfil the need
of the LCA, the outcome must reach the target within the basic range, otherwise, the
redefinition and repetition of LCA may be incomplete. More information and studies
about the LCA policy is mentioned in the literature.

14.4   LCA OF GREEN COMPOSITES

LCA is an important instrument for the manufacture and harvesting of raw compounds. Taking into consideration the timing and investment of resources, LCAs help
determine whether the development of these combinations is profitable. To this end,
LCA is a comparison instrument in the petroleum products and perishable compounds.
Out of the completely synthetic compounds and completely degraded compounds, an
initial petroleum-based natural matrix is made. LCAs utilize a "cradle-to-gate" study,
in which only the production phase is analyzed for comparison with other studies. The
"cradle-to-grave" test is a significant green compound since the end-of-life loss may
differ distinguishably from that of synthetic compounds (Yates et al., 2013).

An important feature of LCA research may be the selection of impact categories in
such a way that shows the purpose and result of the analytical study. The raw compound is usually present in grains yielded from agricultural processes. Hence, most
probably, the LCA gets agricultural impacts of impact which may be studied in global
greenhouse warming, acidulation, eutrophication, ozone depletion, and land use
(Pegoretti et al., 2014; Baum et al., 2020).

14.4.1   Production Phase

This phase usually uses the compounds and resources with which most of the waste
is generated. Both fiber and matrix should be taken individually, in the present step,
because both of them show their presence in the system. For fibers, the product usually
gets growth of plants, their harvesting process and its transportation, and also manufacturing. This requires extraction and manufacturing of raw composites, followed by their
travelling and manufacture. Just as natural fibers, there are several biodegradables which
can be produced from plants. Figure 14.2 shows various types of natural fibers and their
sources. During the production and manufacturing of LCA for green composites, choice

Natural Fibres

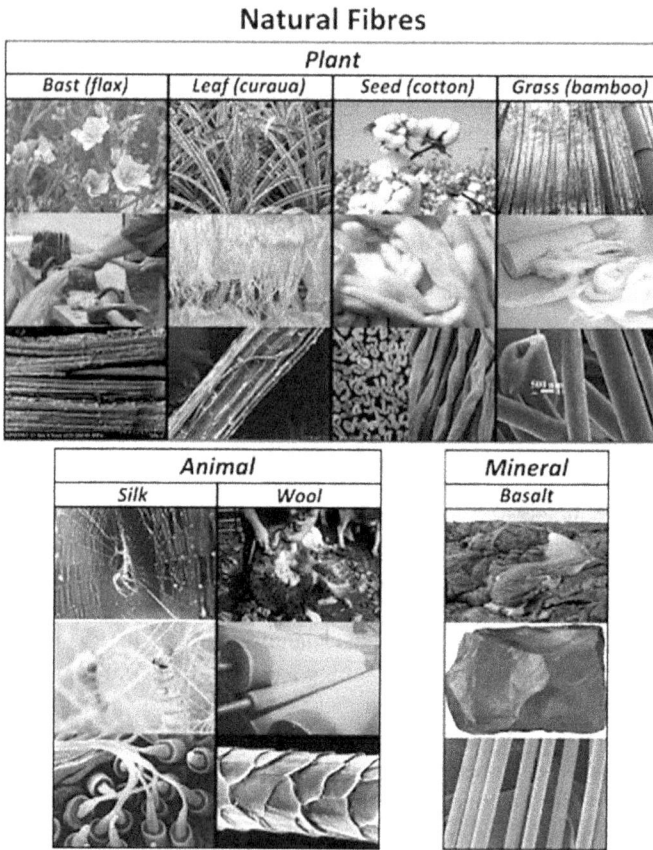

FIGURE 14.2   Different types of natural fibres and their sources.

of the location is significant. Initially, plants must grow in an environment conducive to their growing parameter. Depending upon the maximal growth, from the manufacturing site, the raw composite should be transferred. In addition, the position may be important in calculating an area responding to specific practices. This is why; the use of secondary data may be less accurate than the collection of primary analysis.

Also, land manufacture is required for the manufacture of raw product for green alloys and affects the surrounding phase; LCA must also include land-utilization effects. Natural land is also belonging to a resource because using pitch for the harvesting process decreases the quantity of land. With the change in pitch with the geographical area, there is a change in impact on the pitch, i.e. the place of harvesting process must be taken into account.

The harvesting process of natural fibers includes cultivation and sowing, its process of irrigation, fertilizer, and pest management control, followed by pruning and harvesting process. The environmental effects of many stages of manufacturing can vary much between the types of fiber depending on its need of the plant for maximal production. For example, agriculture generally may not need pesticides or herbicides usage because it will inhibit pests and finish the growth of dangerous fungi. Ploughing and the process of sowing require tools for electricity and the use of fossil fuels, as well as equipment that releases pollutants. Depending upon the limitations of the field, the effect of the machines can also be taken into account (Yates et al., 2013). Farmland often lacks the nutrients needed to grow crops. As a result, the use of fertilizer is significant in eutrophication in the neighbouring ecosystem. Eutrophication is visible when there is a great quantity of nutrients available, and they promote the production of excess biomass in ecosystems. This may give rise to dangerous environmental imbalances. After harvesting, the harvested garbage is transported to the production and harvesting site and transported to deliver the material to the customer. The transport factor is usually given by an estimate of the distance travelled which leads to the amount of gas used (Yates and Barlow, 2013; Brentrup et al., 2004).

The harvesting process of the composite product before final manufacturing process can vary greatly between plant types and intended use. In addition, many tools are used in the manufacturing technology. The determination of inputs and outputs must be focussed in the procedure. Garbage generated by cutting and manufacturing can be used to burn fuel to generate heat for different processes. Various studies (La Rosa et al., 2014) on cork polymer composites, have been published, in which the product was baked with heating process. This procedure is an old method and many surveys have examined the vital uses of crop dump (Maraveas et al., 2020). Many scientists (Kengkhetkit et al., 2014) used pineapple garbage for mixed procedure. This derived good results in reports and rigidity of composition. Nistick et al. showed the formation of composite product from dumped tomato plant (Nisticò et al., 2017). Also, the use of industrial material as energy sources can reduce environmental impact of dangerous and renewable materials.

14.4.2 Use Phase

In general, the usage stage in LCA can be ignored because the composites and materials have very similar usage. Hence, it is very crucial for production and consumption of the same quantity of material during their consumption step (Nair et al., 2021). Glass fiber was compared with LCA and maintained as a natural fiber blend. Composites were made for fittings for the sewage. The utilization phase was excluded as this phase was identical for the two compounds. These predictions are very likely in LCA studies but are many times invalid. In many cases, cost durability has a significant impact on the environment, as a longer service life leads to less material handling and replacement. However, the lack of adequate knowledge about utilization behaviour and degradation limits the total scrutiny of the utilization stage in LCA, for abnormal substances such as green alloys. Use of green alloys is restricted due to several feasibility-related values including humidity, biodegradation, UV exposure, and weather effects (Mayandi et al., 2019; Bari et al., 2019; Joshi et al., 2004; Shogren et al., 2019).

In the automobile unit, green alloys are rather more attractive because they are derived lighter by replacing cloth synthetic fibers with natural cloth fibers, and hence enhancing the fuel efficiency. There are numerous benefits to increasing fuel efficiency in an automobile. Low fuel composition use is financially advantageous as it saves finance on fuel costs over time.

14.4.3 END-OF-LIFE PHASE

For raw natural fiber, division must be given to how the mixture is spoiled for its useful life. Green blends have favourable environmental effects throughout life compared to regular synthetic fiber blends. Composites containing synthetic cloth and non-biodegradable clothes are generally not reusable or cheaply recyclable. At the end of the useful life of all compounds, this will be crucial to get waste-free atmosphere.

Another option to study natural cloth fiber is incineration. Plant crop fibers store carbon during harvest, and are stored for usage lifetime. Lastly, these clothes are burnt for releasing energy in the form of carbon. Baking the original mixture produces carbon dioxide. Meanwhile, burnt carbon is removed from the atmosphere by crop and is collected, with no enhancement in emission of energy.

Another option is to use only biodegradable materials so that at the end of their life, they will decompose naturally without waste. Furthermore, although both the cloth and the fiber undergo biodegradation, they are not required to be separated at the garbage dump. Regulated atmospheric and bacteria or enzymes are required to be decomposed into biodegradable materials. If the raw material is totally compostable, it will naturally break down in the compost pile along with other biodegradable dumps in a favourable temperature and atmospheric condition.

Also natural cloth fibers require dumping and incineration; these disposal processes involve large disposal costs and do not produce new products. Some other alternatives for natural cloth fibers are recycling as well as cutting. For textile raw materials, recycling is identified as the main alternative to the lifetime manufacturing of these raw products (Hawley et al., 2001; Sule et al., 2001; Pensupa et al., 2017). The recycling process may include processes such as up cycling and/or down cycling. Natural cloth fibers are also used to produce recycled products which are of the same or higher or lower quality and value. When they are the same, it is called a recycled product. However, when it is higher or lower, it may be called an upcycled or downcycled product. Innovations in textile waste recycling are expected to lead to vital survey in the future.

14.5   LIFE CYCLE COSTING

We are aware that the LCA pays much attention to the atmospheric problems arising due to the product or system, while the procedure is not effective on the economic impacts of the product's life cycle. Many times, analysis accounts for overall evaluation and effectiveness of the composites. The initial way to achieve the procedure is by conducting a Life Cycle Cost (LCC) study. LCC serves as an instrument to evaluate the full price and benefits of a product, from short-term to long-term. As with LCAs, there is no single, comprehensive procedure to be followed for all LCCs. Instead, there are a number of criteria and guidelines for guiding in designing a workspace and process. In addition, many processes are derived for some particular usage and application in industries. However, while studying the initial stage of LCC, the main aim is to "ensure" that total costs incurred by a product or field are divided into the product process.

There are three classifications of LCCs: traditional, social, and economic. The traditional LCCs have been originally classified for estimation of actual, internal value incurred by the manufacturer or consumer of the product. However, unless the use and disposal phase is the manufacturer's financial responsibility, this alternative is not always addressing the whole life cycle of the raw composite. In addition, the traditional LCC is not focussing on addressing many other reusable materials, as it pays attention on the final value per person. Therefore, the atmospheric LCC was derived to surpass the traditional LCC by incorporating whole life cycle stages and expected costs when affiliated with the LCA. In addition, social LCC estimates the cost that someone in the society will cover complete life cycle. Social LCC is based on atmospheric LCC, with the inclusion of external value evaluation. Social LCC concentrates on the whole people community, including those affected in the future. Since this literature study focuses on the LCA of natural cloth fiber, the atmospheric LCC is presented more clearly.

The main objective of the atmospheric LCC is to use projections of both atmospheric and economic areas to guide raw material manufacturing and make a favourable burden in atmospheric and economic terms. Also, the atmospheric LCC minutely relates to the LCA process; the economy is not governed by the exact aim and restrictions of the LCA production pitch. Literature review should be conducted in addition to the life cycle by calculating single closed boundaries for the pitch. Even, the least important element of LCA is for monetary advantage which may increase and decrease at all steps of life cycle. Therefore, the main flowchart for atmospheric LCC includes the following stages: First is the research and development, next is material production, the next step is manufacture, followed by usage and then the maintenance. Research and development is usually not defined in the LCA as it is significantly not involved in the atmospheric effect. LCI calculates all the initials of the pitch. By enhancing LCI inflows by the particular industry price or neighbouring value, the inflow price can be found. Later on, only the steps in the LCA should be specifically indicated for its LCC (Hunkeler et al., 2008). Care is given when doing LCC as there is a risk of doubling the cost. LCA involves upstream processes where the material is removed and collected. In the LCC, however, when one product composite is removed/cut for making of another material, the price of the product composite can be doubled. For instance, if copper is required, the price of aluminium used in copper can be calculated, but that total value includes the total cost of copper. Therefore, if the value of both copper and aluminium are calculated together, the number of aluminium doubles. The maximum cost to purchase a product (including service charge) should be summed up in the price of that product (including service charge), rather than calculating all the different inputs separately. How is the study conducted? Unlike LCA, the values of each step take into account the manufacturer or consumer perspective. The manufacturer focuses on the cost of the product and the selling price of the product, while the consumer is concerned only about the product value, and disposal costs required during the use of the product (Heijungs et al., 2013).

14.5.1 Durability and Life Predictions

In today's life, researchers and engineers have been asked to explore alternatives in the use of scientific procedure and materials to decrease the overall atmospheric issue. An important innovation to achieve the scope is original blends produced from

TABLE 14.2

Table Showing the Mechanical Properties of Commonly Used Natural and Synthetic Fibers

Fiber	Tensile strength (Megapascal)	Young's modulus (Gigapascal)	Density (gram/cc)	Failure strain (in %)
Abaca	400–813	12–33.6	1.5	2.9
Oil palm	206–248	3.2–3.567	0.7–1.55	4–25
Bamboo	140–230	11–17	0.6–1.1	1.3
Sisal	511–710	9.4–22	1.34–1.5	2–3
Hemp	550–900	68.9–70	1.47–1.48	1.6–4
Curaua	500–1150	11.8	1.4	3.7–4.3
Flax	345–1500	27.6–100	1.5	1.2–3.2
Kenaf	930	53	1.45	1.6
Jute	200–800	13–55	1.3–1.49	1.16–3
Ramie	400–938	24.5–128	1.5	2.5–3.8
Aramid	3000–3600	60–179	1.44	1.9–4.4
Pineapple	170–1627	1.44–82	0.8–1.6	2.4–14.5
E-glass	3400	73	2.55	4.4–4.8
Coir	130–1150	4–6.2	1.2	15–40

atmospherically friendly reusable products. Wide durability, low manufacturing value, biodegradation, and rigidity similar to glass fibers are considered as the basic usage of natural cloth fibers and metals. However, there are some limitations to the feasibility of natural cloth fibers in comparison with synthetic cloth fibers. While assessing the overall atmospheric effect of a mixture, its feasibility plays a crucial role as a deciding parameter to consider for its usage. Degradability of metals, which depends upon the time parameter, can be a constraint on scientists and researchers for correct selection of the compost material for the long-term usage and applications. This is so, as this may sometimes lead to product spoilage along with a significant reduction in mechanical properties. The mechanical properties of commonly used natural and synthetic fibers are shown in Table 14.2.

One way to unify the effects of sustainability on atmospheric effect is to use LCAs. Several atmospheric effect categories can be used to measure the environmental impact of overall sustainability. To continue, the comparison of LCA of composite materials with and without durability results gives rise to the viability of the parameters included in conducting an overall survey. Most studies, including meteorological effects on 26 composites, prefer the use of artificially accelerated environments inside aging rooms rather than the natural environment. Since the regularity cycle, duration, intensity, and exposure conditions of the artificial environment often do not fully reproduce the more unpredictable features of the natural atmosphere, these studies are not directly related to natural weather procedure. In replacement, they deliver an initial application of all negative policies under pre-determined risk cycles.

14.5.2 Moisture Absorption

Moisture intake is a dominant variable in determining the durability of a stated mixture. Moisture absorption is vital in the case of composites made from natural fibers, as these fibers absorb water more easily than hydrophilic and synthetic fibers. The main source for this hydrophilic trend is hemicellulose, a structural component of natural fibers. Moisture intake reduces interfacial fiber-matrix adhesion as an intermolecular hydrogen bond between water and fiber. In addition, swelling of fibers by moisture produces stress at the fiber-matrix interface causing micro-cracking in the capillary promoting matrix. Water-soluble materials then pass through the fibers, forming further bonding between the fibers and the matrix. Finally, water can notably amend the mechanical properties of certain polymers because of the plasticizing effect or hydrolysis of water on the polymer chains and molecules on account of moisture (Holm et al., 2006).

The outcome of moisture absorption in natural fibers significantly reduces mechanical properties and compromises long-term durability. To measure mechanical property deterioration, tension and ductility tests were performed before and after immersion. The results predicted that lower maximum flexural stress could be tolerated by wet specimens. High failure pressure values were noted down for all wet samples, probably due to plasticization or moisture absorption of the fibers as a result of the fiber-matrix interface. However, the tensile test results differ between test specimens containing different types of fiber fractions because a wet specimen with a higher fiber content has a higher tensile strength than powder. The mixtures were dried and tested after obtaining water saturation by immersion for 1400 hours. The mixtures were kept in a container filled with tap water at room temperature. The tensile strength and modulus of water-saturated alloys were observed to decrease by 67%–75% and 74%–83%, respectively, upon drying. Furthermore, compared to the dry and wet samples, the flexural strength and modulus decreased by 57%–73% and 68%–78%, respectively. The high tensile and flexural strains experienced by wet specimens have been suggested by the study authors as a plasticization effect induced by a low cellulose content during water penetration.

Many techniques can be employed to lessen the negative impacts on the properties of natural fiber composite due to moisture absorption. Fiber volume fractionation directly influences the overall longevity and environmental impact in natural blends consisting of natural fibers and synthetic matrixes. The longevity of a mixture can be enhanced by decreasing its fiber volume fraction and, conversely, the hydrophilic character of natural fibers ensuring more moisture absorption resulting in rapid deterioration. However, the use of low fiber content, resulting in increased content of synthetic polymer matrix, thereby, having adverse environmental effects because of the low bio-degradability of synthetic matrices. Another method used to reduce moisture absorption is fiber treatment with chemical compounds. Modifying the fiber with specific compounds reduces the hydrogen bonding capacity of the cellulose by the alkylation process and dissolves the hydrophilic hemicellulose in the natural fiber. The end result is fibers with lower hydrophobicity, better fiber/matrix bonding, and better moisture durability.

14.5.3 Ultraviolet Absorption

Studies have suggested that the extended exposure of the ultraviolet radiation (UVR) upon the natural and synthetic compounds can reduce their durability and life-span.

The photodegradation effects from the exposure of ultraviolet radiation in the natural compounds containing organic compositions are significant. Organic polymers containing covalent bonds break down upon exposure to the ultraviolet radiation, resulting in yellowing, discoloration, surface roughness, brittleness, and overall mechanical property degradation. The tensile strength of fiber-reinforced polymer composites is reduced when the UV radiation gets absorbed in the polymer matrix, thus weakening the matrix and bulking the matrix material. Additionally, the photo-oxidation matrix of the polymer surface propagated by the ultraviolet radiation utilizes ambient oxygen before the process of diffusion. Furthermore, an oxygen gradient is produced due to the degradation concentrated on the surface of the polymer. The oxygen gradient leads to the density gradient, that initiates and expands the cracks that damage the mechanical properties along with the small polymer chains from the chain fragmentation.

Several studies in the published works have explored the effects of ultraviolet radiation on the degradation of composite materials and their mechanical properties. Tensile tests were performed upon the samples using an Instron machine. The largest reduction of tensile strength in any model is found to be 92.57% in the pure (clean) PP models. Samples having 10%, 20%, and 30% fiber weight fractions had decreased tensile strengths of 58%, 37%, and 23%, respectively, while retention increased by increasing the fiber weight fractions in the mixture. Samples were kept for 300 h and 600 h exposure times inside an aging chamber after being irradiated with the ultraviolet radiation, after which the three-point bending tests were performed on the samples. As anticipated, the irradiated specimens for 300 h showed a decrease and a 30% and 41% depletion in the flexural strength and modulus, respectively as compared to the non-radiated specimens. Interestingly, the irradiated models showed a depletion of modulus and flexural strength of 14% and 1.5%, respectively in contrast to non-irradiated specimens for 600 h. It has been proposed that the prolonged exposure time of UV radiation allows for the re-absorption of free radicals resulting, leading to a greater cross-linking between both the neighbouring cells from the fiber and the resin and eventually forming their previous degraded state to the mixture. Mechanical properties increased. These researches suggest an important correlation between material degradation and UVR exposure, but also suggest added complex mechanisms responsible for prevention of degradation in some compounds. Research results have constructed solutions to decrease damage of UVR on the composite materials.

Advocated techniques of engineering the ultraviolet radiation depletion resistance in natural alloys are a hybridization of reinforcement materials in natural alloys and the usage of synthetic plastics as a matrix. Matrix materials reinforced with natural fiber, photo stabilizers, and UV inhibitors may enable reduction of UV damage to fibers, including plastics, with an additional and beneficial increase of water absorption resistance. Research suggests that fibers having small diameters are more suited for UV protection because they have lower UVR transmission. In addition, materials having higher refractive index absorb less UVR, and fibers with a lower porosity also offer higher UV protection. Such parameters while designing composites for UV extreme environments should be taken into account.

14.5.4 BIODEGRADATION

A pivotal parameter in deciding methods and the service life for lifetime dumping for composite materials is the bio-degradation. Enzymatic reactions are distinctive and occur only under definite environmental conditions. Microorganisms such as the bacteria, fungi, and actinomycetes employ enzymes to break down the organic polymers. Cellulose is biodegradable by the fungi and bacteria over the pH (up to 9) and broad range of temperatures (up to 85°C). Different materials have significantly different mechanisms of biodegradation due to the distinctiveness of the enzymes. A few important factors for determining the rate of degradation are crystallinity, molecular weight, crosslinking, structural porosity, and the environmental conditions.

The mechanical properties of alloys are reduced due to the biodegradation during their life-span. The degree of weight loss of the tested mixture can be estimated by the degree of decline that occurs after a certain period of time. The mechanical properties changed after biodegradation. The samples were placed in a compost-filled household waste processing machine for 20 days, tests were conducted on the samples to measure the tensile strength and weight loss at 0–5, 10, 15, and 20 days. The tensile strength of alloys initially dropped from approximately 250-MPa to 50-MPa after a full 20-day period, with a significant decrease in strength noticed within 2–5 days. Over the same duration of 20 days, the degree of degradation of the compounds was greater than 15%, with a sharp increase after 15 days. Biodegradation is deliberated as an environmentally friendly end-of-life disposal technique due to intended life-span and the use of mixture. In a publication by (Ji et al., 2021) tests were conducted on the starch-sisal green composites in combination with organic and inorganic fillers for their biodegradation rates and mechanical properties. The organic filler used is an eggshell powder (EP) that does not contain filler (NF) in the control group. The fillers are blended with the fiber and starch matrix using a blender, and by hot-pressing the resulting slurry to form the final mixture. Biodegradation was carried out by burying the soil for a period of about 30 days, after which the specimens were retrieved and their initial and final masses compared. The NF-composite biodegradation resulted in maximum weight loss of 71%, followed by 67%, 61%, and 60% for EP-, CC- and TP-composites, respectively, excluding the mass of fillers. The NF compound has the highest decay rate because the filler samples are tightly attached to the matrix and prevent microbial decomposition. Tensile and compression tests as well as the water absorption experiments showed better mechanical properties and water resistance from EP-composite compared to the other three. Observing the results, bio-fillers are a viable, inexpensive method for producing biodegradable green composites with high performance.

If the purpose is to reduce the biodegradation of composite material and reduce its life-span, then the usage of biocides should be examined. A biocide, actively employed in industries to protect the materials by destroying the microorganisms that cause bio-degradation, is a broad group of chemical additives. Multiple compound biocides provide greater protection against microbial attack. Other properties of biocides should also be considered within the scope of specificized implementation and environment to augment its potency and inhibit an environment. These properties are temperature, UV stability, solubility, toxicity, working pH in water, and cost-effectiveness.

14.6   CLIMATE TEMPERATURE AND CLIMATE EFFECTS

The meteorological studies found in the text have mixtures of moisture absorption along with the ultraviolet absorption, and biodegradation on the samples of the materials. Although around 30 natural meteorological studies can precisely imitate the state that a substance acquired entirely in its life cycle, these findings are frequently impossible because of the amount of time required to gather valuable data over the years. Regardless, many natural weather studies confirm material degradation and loss of mechanical properties after exposure of the material to natural environments.

Artificial or accelerated weathering is a procedure happening in the aging rooms that "attempts to mimic the harmful effects of the natural environment and long-term external exposure via exposing the test specimens to ultraviolet radiations, heat and moisture in a supervised manner." Variable humidity, UV, and temperature conditions may be set in stages to reproduce environmental aging. However, the results related to the actual conditions due to the regularity, exposure conditions, duration, and intensity of the cycles of rapid weathering may not be accurate.

14.7   CONSISTENCY OBSERVATIONS IN LIFE CYCLE ASSESSMENT

Researchers often assume cradle-to-grave analysis while examining the environmental effects of bioplastics and bio-based composites, ignoring the use-stage of the material over its lifetime. Shortcomings to the information, exceptionally for the novel materials like the green composites about the behavioural and degrading properties in service can impede the assessment of the environmental impacts of physical service life. Therefore, service life performance must be integrated with the LCA and material design processes to come up with a thorough assessment of environmental impact.

The LCIA listings reviewed in the study were "Carbon feedstock for resource utilization and polymers during wood cultivation; material and energy flow during biosynthetic or synthetic polymerization; wood by-products and consequent polymer refining processes." Although supplementary LCAs were also created based on the amount of material required to fulfil the design standards. Environmental effects vary significantly between LCAs. Ranges are slightly higher than 20% when degradation is not considered. In addition, minor differences were observed in all four environmental groups except LCA and degradation for WPC tested in Phoenix Arizona having dry conditions, as less material needs to be replaced in this environment. Therefore, the study summarized that service conditions have the prospective to influence the environmental impact of the material provided, so utilization-phase stability integration within the LCA is essential.

Some important factors such as fiber cultivation requirements, terrain, mechanical properties, and cost are considered while selecting natural fibers for engineering applications (Lakshmi Narayana et al., 2021; Pujari, 2013; Väisänen et al., 2017; Karus et al., 2002; Song et al., 2009). Knowledge of fiber plant utility in many parts of the world can help in the application of plant by-products to reduce waste. In addition to cultivation needs, industrial and economic factors affect the fibers grown in particular countries. Favourable biopolymer properties for composite materials include biodegradability, biocompatibility, weather resistance, and hydrophobicity (Aaliya et al., 2021).

14.8   COST OF FIBER MATERIAL

An interesting attribute about the natural fibers in engineering applications is their economical cost compared to synthetic fibers. Natural fibers are cheap but meet engineering design standards in many cases. Of course, synthetic fibers have better mechanical properties. The specific strength and modulus of natural fibers are similar to those of the synthetic fibers when the density is not calculated. The lower concentrations found in natural fibers is in agreement to lower transport costs, thereby decreasing the price of natural fibers. Consumption-phase cost efficiency of natural fiber has emerged through reduction of weight of composite components in the automobile industry that allows for higher efficiency of fuel. Natural fiber production also requires low energy demand. The low-priced expense of natural fibers means a lower cost for composites reinforced with natural fibers than for synthetic fibers, i.e., the total cost is lower if the price of the fiber and matrix material is different from the amount of natural fiber. Injection moulding processes are expensive in an industry already designed for glass fiber, so glass fiber is the standard. Therefore, other production costs should as well be considered for a more elaborate approach for increasing the cost-effectiveness of natural fibers in industrial settings even if their raw material costs are low.

14.8.1   BIOPOLYMER MATRICES

Biopolymers made from renewable sources have recently aroused interest from the scientific and industrial sectors as a better alternative to the synthetic polymers that are derived from petroleum. Environmental distress of petrochemical plastics, such as global warming, finite fossil fuel availability, rapid landfill deposits, and low biodegradability have led to transferring the research interests in biopolymers to diminish such repercussions. Two separate criterions were used to define the expression "biopolymer" i.e., the source of the raw material and the biodegradability of the polymer. Consequently, biodegradable polymers produced from renewable resources, biodegradable polymers produced of fossil fuels, and biodegradable polymers produced from stable raw polymers are all biodegradable. Bio-based biopolymers are produced by plants, animals, and microorganisms (biological systems). These can also be synthesized by the chemical synthesis of natural materials like starch, sugar, and corn. Furthermore, there are three main classes of natural polymers based upon their structure namely the polysaccharides, polynucleotides, and polypeptides.

A number of biopolymers these days are derived via first-generation feedstock containing consumable biomass (vegetable oils, starch, sugar, etc.) as well as disposable resources like the natural rubber. Cellulose, which is found in cell walls of plants, is a polysaccharide and is the most abundantly found biopolymer. Approximately 10–11 tons of cellulose is produced by plants per year. Other organisms such as bacteria and fungi can also give rise to cellulose. Starch stored as energy in plants is an abundant plant-based biopolymer. Carbs can be extracted from many different sources in nature such as rice, wheat, corn, and potatoes because it's a natural carbohydrate polymer. Biopolymer matrices can be applied in bio-composite materials. They are derived from regenerative materials such as starch, soy, and cellulosic plastics. A simple biopolymer known as Polylactic acid (PLA) is made by the lactic acid polymerization from natural sources such as corn-starch or sugar cane. Collagen is an animal-based biopolymer with most vital resources being pig skins, bovine skins, and beef bones. Polyhydroxyal-kanoates (PHAs) are also biopolymers produced by microorganisms synthesized from regenerative materials by fermentation. Although polycaprolactone (PCL) is a synthetic polyester, it is easily biodegradable by enzymes and fungi (Vroman et al., 2009).

Selection criterions for the biopolymers differ vastly between different industrial applications. The properties considered include pH, density, diffusion coefficient, tractability, durability, melting temperature, refractive index, geological properties etc. Chitosan is a polysaccharides are chosen for coating fruits and vegetables due to their low toxicity, and high biodegradability, along with their anti-fungal, anti-oxidant, and film-forming capabilities. High-level biodegradability and resistance to weather makes PLA compatible with respect to almost 50 composite materials. Green matrix material compounds such as hyaluronic acid-based on 83 biopolymers have been used in medical applications for drug administration. This is because of the fact that they have excellent biodegradability, biocompatibility, and non-toxic properties. In the automotive industries, Mitsubishi Motors uses polybutylene succinate (PBS) and PLA. For all around durability, the compounds made from hydrophilic biopolymers such as starch and cellulose absorb more added moisture than composites produced from hydrophobic biopolymers like PHB and PHBV (both types of PHA). Matrixes for biopolymer composites should be selected primarily on the basis of application-specific criteria, due to the various advantages and characteristics available in their many variants (Fitzgerald et al., 2021).

14.9   CONCLUSION

Summarizing the preceding review, it is safe to say that green blends have demonstrated to be a feasible alternative instead of synthetic blends in various applications. Cost-effective manufacture, outspread availability, low impact on the environment, high specific strength, and hardness are just a few advantages of green alloys over the conventional carbon, glass, Kevlar, and other man-made alloys. LCAs are a scientific tool helping to measure the environmental impact of a product by its production, usage, and end-of-life stages. LCAs can be applied to a wide range of products in different sectors and industries, enunciating that the regulations for managing LCAs do not describe the process of performing each step of the analysis. The LCC study should be used while evaluating the momentary and long-term financial costs and benefits of a system. For instance, just like the LCA, the LCC is a comprehensive analytical tool unescorted by some specific steps to perform all analyses since it focusses to consolidate cost over its life cycle in the decision-making process in the premature steps of development. Therefore, both the LCA and LCC tools should be employed when calculating the economic and environmental implications for an effective use of green alloys.

Ideally, phase stability in green alloys should be included in the LCA studies, but seldom occur in practice as a consequence of bounded information of the material degradation processes. Nevertheless, four separate approaches have been associated in compromising the mechanical properties of green alloys. These are moisture absorption, ultraviolet radiation exposure, biodegradation, temperature and weathering effects. The structural integrity of the mixture is compromised because of the hydrophilic nature of natural fibers causing the fibers to swell because of the absorption of moisture. The ultraviolet radiations incapacitate the composites by disturbing the covalent bonds in the polymer matrix on the polymer surface by promoting photo-oxidation. Biodegradation occurs when the microorganisms break down organic polymers through enzymatic reactions, although biodegradation is often considered an environmentally friendly tool for lifelong disposal. The combination of these three mechanisms with additional temperature effects increasing moisture-induced erosion is measured in natural and rapid meteorological studies.

The cultivation conditions, terrain, mechanical properties, utility, and price are some of the parameters to be considered while selecting reinforced natural fibers in green alloys. Different fibers have different requirements for optimum rainfall, temperature, and soil conditions. The geographical availability of certain fiber kinds may also limit the type of fibers that can be used in a particular area. Understanding the mechanical properties of natural fibers for a particular application despite the fact that there are significant variations in the reported mechanical properties of natural fibers because of inconsistent testing methods can help selecting the appropriate type of fiber. Preferably, natural fibers should be obtained from plants that have by-products from non-fiber replicas to reduce plant waste. The price of different types of natural fibers varies, but for the most part, they are much less expensive than their traditional synthetic alternatives. Synthetic polymers such as biopolymer matrices in green composites are another viable alternative to support sustainable growth. The selection criteria for green alloys are highly dependent on weather and moisture resistance. Other criterion also includes the biocompatibility and biodegradability.

Elaborate studies have been performed on the life cycle analysis (LCA), durability, and mechanical properties of green alloys showing that they work from these categories since they belong to green knitted alloys. Further studies on the mechanical properties of the green composite must look into the effects of multiple composite parameters, together with the linear yarn density, volume fraction of the fiber, a wide range of knitting angles, and various compositions of the fiber matrix for optimal composite characterization. Minute or no data is available for the life cycle analysis of green knitted alloys. LCA and LCC studies should be conducted to assess their ability to meet cost requirements, specifically in contrast to synthetic composites with conventional structures. Sustainability considerations should also be included in these analyses for more accurate assessment of environmental implications. Climate studies on green knitted alloys should be conducted to assess their phase-use longevity and durability, as well as biodegradation inspections for the development of procedures for lifetime disposal.

REFERENCES

Aaliya, B., Sunooj, K. V., & Lackner, M. (2021). Biopolymer composites: A review. *International Journal of Biobased Plastics*, *3*(1), 40–84. https://doi.org/10.1080/24759651.2021.1881214

Bari, E., Morrell, J. J., & Sistani, A. (2019). Durability of natural/synthetic/biomass fiber-based polymeric composites: Laboratory and field tests. In Jawaid, M., Thariq, M., & Saba, N. (Eds.), *Durability and Life Prediction in Biocomposites, Fibre-reinforced Composites and Hybrid Composites* (pp. 15–26). Cambridge: Woodhead Publishing.

Baum, R., & Bieakowski, J. (2020). Eco-efficiency in measuring the sustainable production of agricultural crops. *Sustainability*. https://doi.org/10.3390/su12041418.

Brentrup, F., Küsters, J., Lammel, J., et al. (2004). Environmental impact assessment of agricultural production systems using the life cycle assessment (LCA) methodology II. The application to N fertilizer use in winter wheat production systems. *European Journal of Agronomy, 20*, 265–279.

Curran, M. A. (2016). *Life Cycle Assessment* (pp. 1–28). Kirk-Othmer Encyclopedia of Chemical Technology.

Fitzgerald, A., Proud, W., Kandemir, A., Murphy, R. J., Jesson, D. A., Trask, R. S., Hamerton, I., & Longana, M. L. (2021). A life cycle engineering perspective on biocomposites as a solution for a sustainable recovery. *Sustainability, 13*, 1160. https://doi.org/10.3390/su13031160

Gupta, A., & Kumar, A. (2010). Composites materials: Addressing the climate change. *Asia Pacific Business Review, 6*(1), 78–89. https://doi.org/10.1177/097324701000600107

Hawley, J. (2001). Textile recycling as a system: The micro-macro analysis. *Family and Consumer Sciences Research Journal, 92*, 40–46.

Heijungs, R., Settanni, E., & Guinée, J. (2013). Toward a computational structure for life cycle sustainability analysis: Unifying LCA and LCC. *The International Journal of Life Cycle Assessment, 18*, 1722–1733.

Holm, V. K., Ndoni, S., & Risbo, J. (2006). The stability of poly(lactic acid) packaging films as influenced by humidity and temperature. *Journal of Food Science, 71*, E40–E44.

Hunkeler, D., Lichtenvort, K., & Rebitzer, G. (2008). Introduction: History of life cycle costing, its categorization, and its basic framework. In *Environmental Life Cycle Costing* (pp. 35–50). Boca Raton, FL: CRC Press.

ISO. (2006a). *Environmental Management—Life Cycle Assessment—Principles and Framework*. Geneva. www.iso.org/standard/37456.html

ISO. (2006b). *Environmental Management—Life Cycle Assessment—Requirements and Guidelines*. Geneva. www.iso.org/standard/38498.html

Ji, M., Li, F., Li, J., et al. (2021). Enhanced mechanical properties, water resistance, thermal stability, and biodegradation of the starch-sisal fibre composites with various fillers. *Materials & Design, 198*, 109373.

Joshi, S. V., Drzal, L. T., Mohanty, A. K., et al. (2004). Are natural fiber composites environmentally superior to glass fiber reinforced composites? In *Composites Part A: Applied Science and Manufacturing* (pp. 371–376). London: Elsevier.

Karus, M., & Kaup, M. (2002). Natural fibres in the european automotive industry. *Journal of Industrial Hemp, 7*(1), 119–131.

Kengkhetkit, N., & Amornsakchai, T. (2014). A new approach to 'greening' plastic composites using pineapple leaf waste for performance and cost-effectiveness. *Materials & Design, 55*, 292–299.

La Rosa, A. D., Cozzo, G., Latteri, A., Recca, A., Björklund, A., Parrinello, E., Cicala, G. (2013). Life cycle assessment of a novel hybrid glass-hemp/thermoset composite. *Journal of Cleaner Production, 44*, 69–76. https://doi.org/10.1016/j.jclepro.2012.11.038

La Rosa, A. D., Recca, G., Summerscales, J., et al. (2014). Bio-based versus traditional polymer composites. A life cycle assessment perspective. *Journal of Cleaner Production, 74*, 135–144.

Lakshmi Narayana, V., & Bhaskara Rao, L. (2021). A brief review on the effect of alkali treatment on mechanical properties of various natural fiber reinforced polymer composites. *Materials Today: Proceedings, 44*, 1988–1994.

Mansor, M. R., Mastura, M. T., Sapuan, S. M., et al. (2018). The environmental impact of natural fiber composites through life cycle assessment analysis. In Jawaid, M., Thariq, M., & Saba, N. (Eds.), *Durability and Life Prediction in Biocomposites, Fibre-reinforced Composites and Hybrid Composites* (pp. 257–285). Cambridge: Woodhead Publishing.

Mansor, M. R., Salit, M. S., Zainudin, E. S., et al. (2015). Life cycle assessment of natural fiber polymer composites. In Hakeem, K., Jawaid, M., & Alothman, O. Y. (Eds.), *Agricultural Biomass Based Potential Materials* (pp. 121–141). Cham: Springer International Publishing.

Maraveas, C. (2020). Production of sustainable and biodegradable polymers from agricultural waste. *Polymers*, *12*, 1127.

Mayandi, K., Rajini, N., Manojprabhakar, M., et al. (2019). Recent studies on durability of natural/synthetic fiber reinforced hybrid polymer composites. In Jawaid, M., Thariq, M., & Saba, N. (Eds.), *Durability and Life Prediction in Biocomposites, Fiber-reinforced Composites and Hybrid Composites* (pp. 1–13). Cambridge: Woodhead Publishing.

Nair, A. B., Sivasubramanian, P., Balakrishnan, P., et al. (2022). Environmental effects, biodegradation, and life cycle analysis of fully biodegradable 'green' composites. In Thomas, S., Joseph, K., Malhotra, S. K., et al., (Eds.), *Polymer Composites, Biocomposites* (pp. 515–568). Hoboken, NJ: Wiley Blackwell.

Netravali, A. N., Chabba, S. (2003). Composites get greener, *Materials Today*, *6*(4), 22–29. https://doi.org/10.1016/S1369-7021(03)00427-9

Nisticò, R., Evon, P., Labonne, L., et al. (2017). Post-harvest tomato plants and urban food wastes for manufacturing plastic films. *Journal of Cleaner Production*, *167*, 68–74.

Pegoretti, T. D. S., Mathieux, F., Evrard, D., et al. (2014). Use of recycled natural fibres in industrial products: A comparative LCA case study on acoustic components in the Brazilian automotive sector. *Resources, Conservation and Recycling*, *84*, 1–14.

Pensupa, N., Leu, S. Y., Hu, Y., et al. (2017). Recent trends in sustainable textile waste recycling methods: Current situation and future prospects. *Topics in Current Chemistry*, *375*, 1–40.

Pujari, S. (2013). Comparison of jute and banana fiber composites: A review. *International Journal of Current Engineering and Technology*, *2*, 121–126. https://doi.org/10.14741/ijcet/spl.2.2014.22

Ramesh, M., Deepa, C., Kumar, L. R., et al. (2020). Life-cycle and environmental impact assessments on the processing of plant fibres and its bio-composites: A critical review. *Journal of Industrial Textiles*. https://doi.org/10.1177/1528083720924730

Rebitzer, G., Ekvall, T., Frischknecht, R., et al. (2004). Life cycle assessment Part 1: framework, goal and scope definition, inventory analysis, and applications. *Environment International*, *30*, 701–720.

Song, Y. S., Youn, J. R., & Gutowski, T. G. (2009). Life cycle energy analysis of fiber-reinforced composites. *Composites Part A: Applied Science and Manufacturing*, *40*, 1257–1265. https://doi.org/10.1016/j.compositesa.2009.05.020

Shogren, R., Wood, D., Orts, W., et al. (2019). Plant-based materials and transitioning to a circular economy. *Sustainable Consumption and Production*, *19*, 194–215.

Sule, A. D., & Bardhan, M. K. (2001). Recycling of textile waste for environment protection-An overview of some practical cases in the textile industry. *Indian Journal of Fibre & Textile Research*, *26*, 223–232.

Thankachen, R. U., Nair, A., Raj, J., et al. (2021). Methodologies for selecting biopolymers and their characteristic features for industrial applications. In Thomas, S., Gopi, S., & Amalraj, A. (Eds.), *Biopolymers, and Their Industrial Applications* (pp. 81–103). Amsterdam: Elsevier.

Tukker, A. (2000). Life cycle assessment as a tool in environmental impact assessment. *Environmental Impact Assessment Review*, *20*, 435–456.

Väisänen, T., Das, O., & Tomppo, L. (2017). A review on new bio-based constituents for natural fiber-polymer composites. *Journal of Cleaner Production*, *149*, 582–596.

Vroman, I., & Tighzert, L. (2009). Biodegradable polymers. *Materials*, *2*(2), 307–344. https://doi.org/10.3390/ma2020307

Yates, M. R., & Barlow, C. Y. (2013). Life cycle assessments of biodegradable, commercial biopolymers—A critical review. *Resources, Conservation and Recycling*, *78*, 54–66.

Index

For Product Safety Concerns and Information please contact our EU
representative GPSR@taylorandfrancis.com
Taylor & Francis Verlag GmbH, Kaufingerstraße 24, 80331 München, Germany

www.ingramcontent.com/pod-product-compliance
Lightning Source LLC
Chambersburg PA
CBHW060555220326
41598CB00024B/3107